IMAGES
of America

GREATER HARTFORD FIREFIGHTING

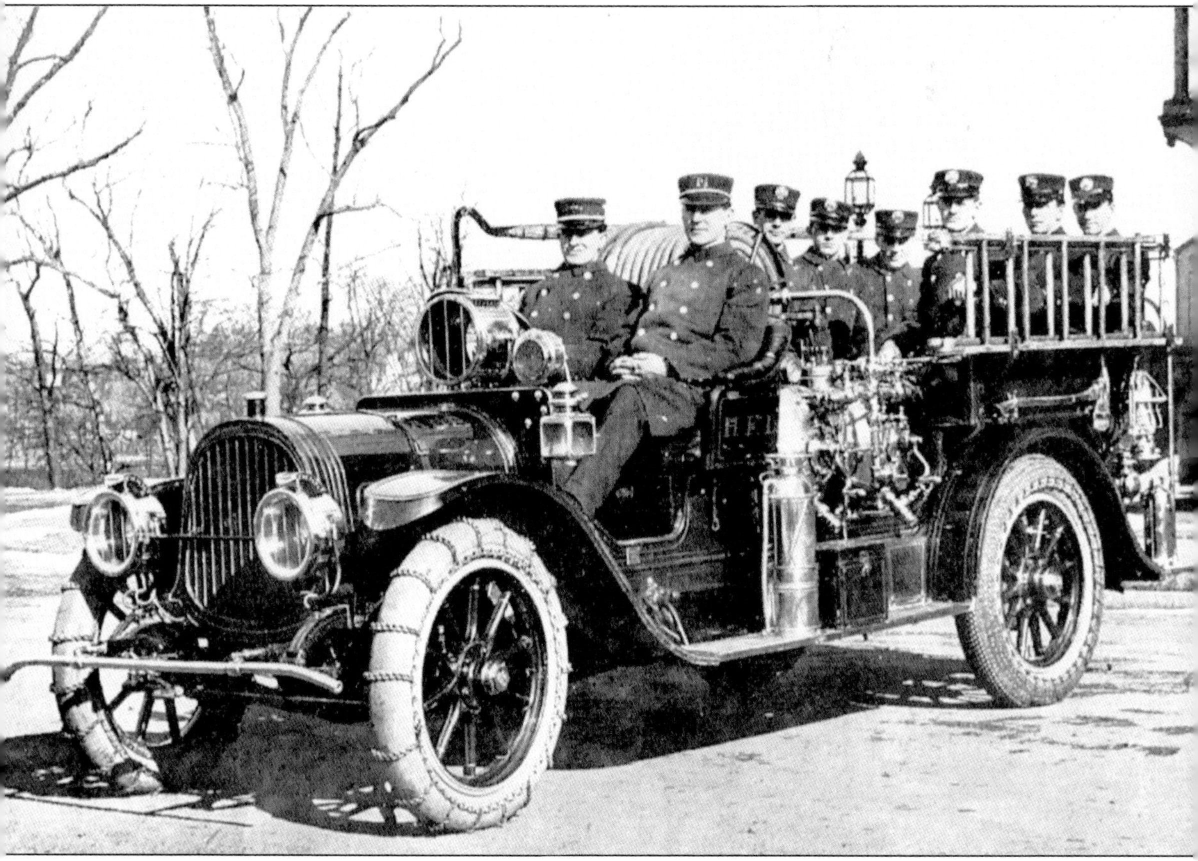

In 1910, the Hartford Fire Department purchased this chemical and hose wagon for Squad A, shown here in front of the state capitol. Located in Hartford, the Pope-Hartford Automobile Company had a reputation for producing a solidly built vehicle. Even so, annual production never exceeded 762 vehicles. They produced touring cars, trucks, and fire apparatus, all built using the same chassis. The company operated from 1904 to 1914. Founded by Col. Albert A. Pope, his manufacturing methods pointed the way for Henry Ford through precision machining and interchangeability of parts and vertical integration. (Courtesy of the Connecticut Fire Museum Archives.)

On the cover: Please see page 46. (Courtesy of Mike Gorman.)

IMAGES
of America

GREATER HARTFORD FIREFIGHTING

The Connecticut Fire Museum

ARCADIA
PUBLISHING

ISBN 978-0-7385-4538-7

Published by Arcadia Publishing
Charleston, South Carolina

Printed in the United States of America

Library of Congress Catalog Card Number: 2006923206

For all general information contact Arcadia Publishing at:
Telephone 843-853-2070
Fax 843-853-0044
E-mail sales@arcadiapublishing.com
For customer service and orders:
Toll-Free 1-888-313-2665

Visit us on the Internet at www.arcadiapublishing.com

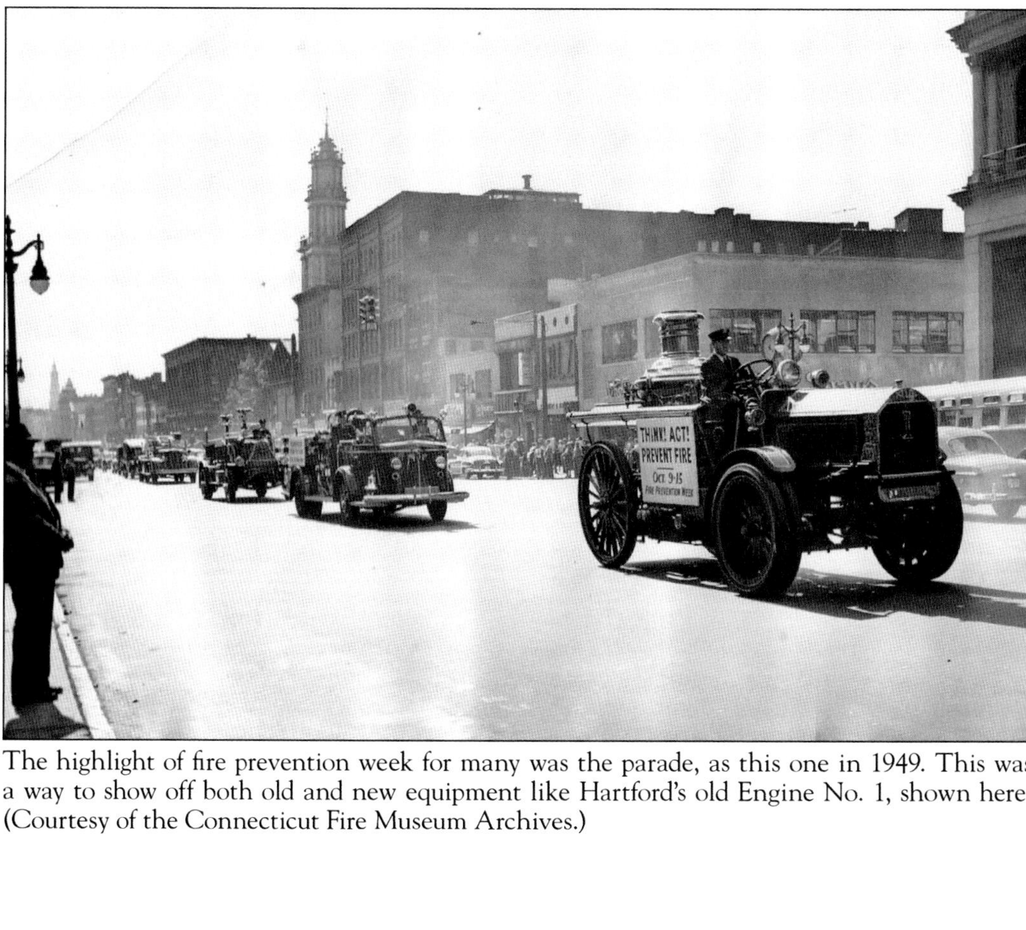

The highlight of fire prevention week for many was the parade, as this one in 1949. This was a way to show off both old and new equipment like Hartford's old Engine No. 1, shown here. (Courtesy of the Connecticut Fire Museum Archives.)

Contents

Acknowledgments

This book would not have been possible without the assistance of retired Hartford firefighter Keith B. Victor, editor of the *Alarm Room News*, and retired Hartford deputy chief Mike Gorman, both of whom shared their collection of Hartford firefighting photographs. Keith encouraged the authors to include Hartford's two neighboring towns, which have cooperated with each other for over 80 years of mutual aid fire protection.

We would also like to thank Thomas E. Holcombe of Burlington, Connecticut; Assistant Chief Gary Allyn of the West Hartford Fire Department; retired West Hartford lieutenant Rudy Desroches of Windsor Locks; and East Hartford Fire Department members Capt. Kenneth Beliveau, Lt. Robert Fournier, firefighter David Prior, Chief Michael Eremita, and retired East Hartford firefighter Jack Griffin, all of whom provided information and photographs for this book.

Bert D. Johanson
President
Connecticut Fire Museum
East Windsor, Connecticut 06088
Telephone 860-623-4732
http://www.ceraweb.org/Firemuseum

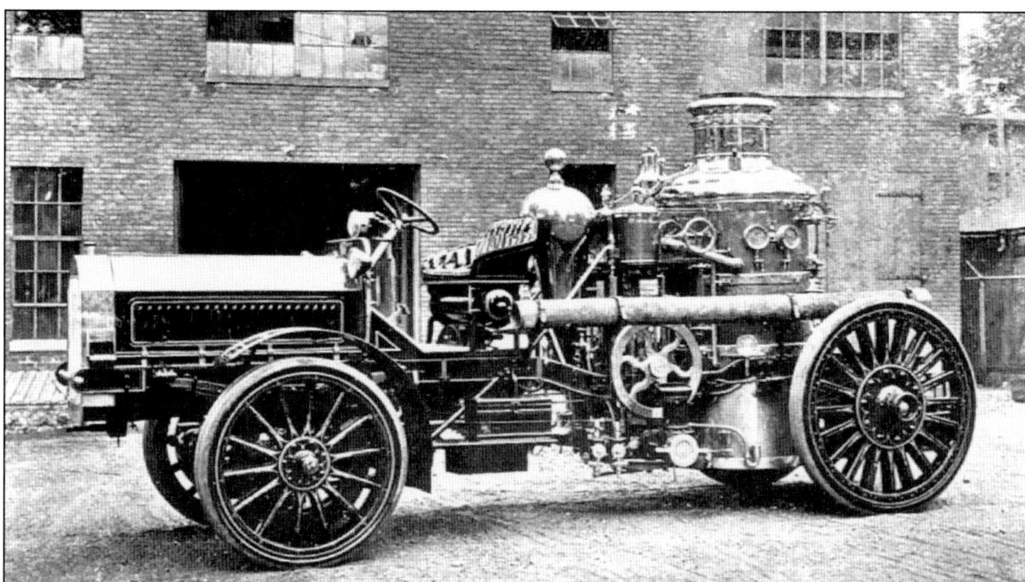

Hartford Engine Company No. 1's 1912 American LaFrance steam pumper is seen here. For a short time, some of the companies manufactured combination combustion engine/steam pumper units. This engine was in regular service until 1930. The Hartford Fire Department donated this apparatus to the Connecticut Historical Society in the 1990s. After a five-year restoration project, it is now on display at the Old State House in Hartford. (Courtesy of the Connecticut Fire Museum Archives.)

INTRODUCTION

Hartford, steeped in history and tradition, is one of the oldest towns in New England and, since its settlement in 1635, has been the business and political center of Connecticut. By the 1890s, Hartford was the center of an agricultural and manufacturing community with steam train and trolley car connections in all directions, affording Hartford businesses a vast market for their goods and services. Hartford was a city of wealth with numerous banks and insurance companies. East Hartford and West Hartford were both once part of colonial Hartford. East Hartford separated from Hartford in 1783 and West Hartford in 1854. Both towns were primarily rural until the 1940s but in 2006 are now residential suburbs of Hartford that have substantial commercial and industrial development.

At the beginning of the 20th century, most vehicles, including fire equipment, were still horse drawn. The use of steam power in firefighting equipment was just dawning. The steam was used to operate the pump; the horses were used to transport this pump to the site. As with steam locomotives, the fire in the steam pumpers needed to be tended 24 hours a day, which lead to a transition from all-volunteer fire departments to paid departments.

Early volunteer fire departments were associated with rivalry and rowdiness, most often associated with New York City's volunteer firemen.

Much of the pride and professionalism felt by firefighters today stems from the actions and traditions of these early firefighters. Traditionally the only essential community service that has been provided solely by volunteers is firefighting. The pride in equipment was born with the early hand engines, used by these volunteer fire departments, that were lavish and elegant, often painted with gold leaf. Anyone who has attended a fire parade recently has seen firsthand that this tradition continues. The language of firefighting today also reflects the history of the volunteer companies. "Making a run," a term used today, comes from the days when pulling a hand-drawn pumper with ropes was the way that equipment was transported to a fire. Even though the technology of equipment has progressed, the language often has not.

Movement from horse-drawn to motorized equipment began during the early part of the 20th century. Many cities had replaced the horse completely by the 1920s. The 1920s saw what may be the most widespread accepted transition to motorized rigs, from cars to commercial trucks to fire engines. Advances in technology provided advancement in motorized vehicles; hydraulics replaced mechanics in things such as brakes and devices for raising aerial ladders.

World War II produced many changes. Industry geared up to supply the troops fighting overseas. The orders for industries in this country were put on hold, and the majority of trucks built by those companies that traditionally made domestic firefighting equipment—American LaFrance, Mack, and others—were sent to the troops instead. At the end of the war, equipment was once again able to be built for the cities, and apparatus overdue for retirement could be replaced.

As the cities grew, the equipment used by firefighters had to grow. Ladder trucks, once equipped with only short, wooden ladders, went to aerial ladders. These longer ladders necessitated longer trucks. Longer trucks were difficult to steer, particularly in crowded city streets, which lead to a "tiller" being added to the rear axel of the truck, from which the rear of the truck could be steered also. After World War II, other things were added, such as a snorkel with an elevated firefighting platform.

Movement to steam equipment also signaled the movement from volunteer to paid fire departments. Some of the larger cities in the country made this move earlier than smaller communities. Cincinnati, Ohio, had the first professional paid department in the United States in 1853. Following their example, Providence, Rhode Island, made the move in 1854; New York City in 1865; Albany, New York, in 1867; and Philadelphia in 1871.

The Hartford, East Hartford, and West Hartford Fire Departments have a proud and rich tradition of service to the community. The membership of the Connecticut Fire Museum, in East Windsor, Connecticut, felt that the history of these departments, as well as the historical photographs in the museum archives, should be shared with future generations to help them fully appreciate the dedication and heritage of the fire service in these towns and cities.

The Connecticut Fire Museum was founded in 1968 by a group of Connecticut Trolley Museum members who found they had a common interest in fire engines as well as streetcars. In the 1960s, the Connecticut Trolley Museum was running a steam locomotive as a supplement to its streetcar operations. Some of the members felt that having a piece of fire apparatus on hand to provide water to the steam engine, and also help the local fire department with the numerous grass fires the engine started, would be a good idea. What started out with one 1930 Maxim fire engine in 1968 had become a collection of over 20 pieces of antique fire apparatus by 1974. The museum opened to the public in 1975 after erecting an 11,000-square-foot building on the Connecticut Trolley Museum's property. Today, in conjunction with the Connecticut Trolley Museum, the Connecticut Fire Museum has over 25,000 visitors per year.

One

THE VOLUNTEER YEARS

The first settlers arrived in the area in 1635. In 1669, the Connecticut General Court ordered that every household be required to keep a fire ladder and bucket in readiness; this was the first step in the progress of fire protection in Hartford. This constituted an early volunteer fire department.

In 1783, the roof of the wooden statehouse caught on fire. Through the efforts of the townspeople, it was extinguished before the structure was badly damaged. The year 1785 saw both the purchase of the first "fire engine" and the beginning of a public water supply system, which an informal group called the Proprietors of the Hartford Aqueduct set up. They placed several hundred feet of wooden waterline near the statehouse. The first "fire engine" was actually a "hand engine." A hand engine was a pumper that the firemen pulled with ropes. A pumper usually had a water tank in the center with a piston pump system that was powered by men on either side of the tank pushing the long handles called brakes. Ladder trucks typically carried several wooden ladders to the site of the fire. Since there were few buildings that were more than two stories tall at that time, they did not need the tall ladders now associated with firefighting.

The first fire safety ordinance in Hartford was passed in 1788, outlawing bonfires on the streets. The organization that would eventually become the Hartford Fire Department was organized in 1789. On February 23 of that year, the town appropriated 300 pounds sterling to "buy engines and build cisterns." In November 1789, the common council appointed one engineer and three foremen to oversee the fire department. Miles Beach, a silversmith, was the first chief of the volunteer department. The council also voted to offer a premium of $10 to the fire company that arrived first at the scene of a fire. Prior to this, the Hartford insurance companies paid cash rewards to the fire company that arrived first at the scene. During this time, Hartford's population had grown to about 5,000, with the city entirely within a one-mile radius of the statehouse.

In January 1812, the first hook and ladder company was organized. This was followed by the Hylas Sack and Bucket Company, organized in 1816 and equipped with leather buckets for firefighting and sacks for removing valuables from the burning property. In September 1827, the first hose company was organized with Nathan Ruggles as its foreman. April 1827 saw a reorganization of the department. Prior to that time, the fire warden was in charge of the department at all fires within the city. At a city council meeting, ordinance was passed by which the chief engineer was given control of the department at fires. For a salary of $50 a year, he was expected to perform these duties in addition to his current duties of "keeping the engines, pump brakes and boxes in repair." The last chief engineer to be appointed by the common council was William Hayden, on October 14, 1833. On April 25, 1842, there was an ordinance passed in Hartford by which the firemen elected their own chief. On July 6, 1842, his fellow firemen elected William Hayden chief.

The common council, in April 1834, appointed a committee "to inquire into the wants of the fire department with a view to its reorganization." Upon the report of the committee, the council ordered 1,600 feet of hose, the first riveted hose used in the city. Each company was voted to receive "$3 when its services were required at fires." Each company was allotted men to fill both the firefighter and officer needs of the company as follows: Engines Companies Nos. 1, 2, 3, 4, 6, and 7 were allotted 45 men each; Engine Company No. 5 was allotted 55 men; the Hook and Ladder Company was allotted 60 men; and the Hose Company was allotted 15 men. The committee recommendations also lead to the purchase of a new engine in May 1834. James Smith of New York, who built many of the fire engines used in other large cities, built this engine, which was ordered for Engine Company No. 4. The old engine was then sold to the town of Wethersfield.

In 1840, the city had seven engine companies, one hose company, and one hook and ladder company, which were manned by about 500 volunteers. Most of the engines were Smith machines, built in New York City. Another hose company, Pioneer Hose Company No. 2, was organized on August 5, 1845, with Ivory H. Blood as foreman. The company occupied a barn at 12 Temple Street.

The department continued to use hand engines until 1861, when the first steam fire engine, Phoenix No. 3, was purchased. From that time on, the hand machines were doomed, and within three years, they were out of service in the city. The first steam fire engine was built in England in 1829 by George Braithwaithe. The first American steam fire engine built for use by a city was in 1852, when Moses Latta built an engine for the city of Cincinnati. Latta was able to design and build a machine that was powerful enough to throw a forceful stream of water yet still portable enough to be brought to the fire. Initially volunteers in the United States were opposed to the new machines. In 1859, a steam fire engine made by Amoskeag of Manchester, New Hampshire, was brought to Hartford for testing. The engine did not meet with the approval of chief engineer Samuel H. Havens, who had an unbounded faith in the just-introduced water system fed by the new reservoir in West Hartford. Chief Havens felt that with the availability of water the city did not need a steam fire engine, but the city council and fire department members overruled him.

As with all cities, Hartford had a number of large, costly fires in its history. Some of the notable fires during the volunteer years included the Fales and Gray Car Works on Potter Street, in March 1854, an explosion and fire that caused the death of 21 people and serious injury to 50 others. This fire resulted in the establishment of Hartford Hospital, as the city did not have a professional health care facility for the injured. The Colt Firearms factory building caught fire on February 1, 1864, resulting in one fatality and $1.5 million in damage. This fire prompted the city to consider a paid fire department.

On October 1, 1864, the common council passed an ordinance that established a paid fire department for Hartford overseen by a board of fire commissioners. The new organization, made necessary also in part by the introduction of the steam fire engine, which required fewer men, went into effect at noon on December 1, 1864. The advent of steam fire pumpers meant the fire service in general was going the route of having more professional departments. Between steam boilers needing personnel to assure that the engine boiler is kept hot at all times and the care and feeding of the fire horses, the job just did not lend itself to volunteers.

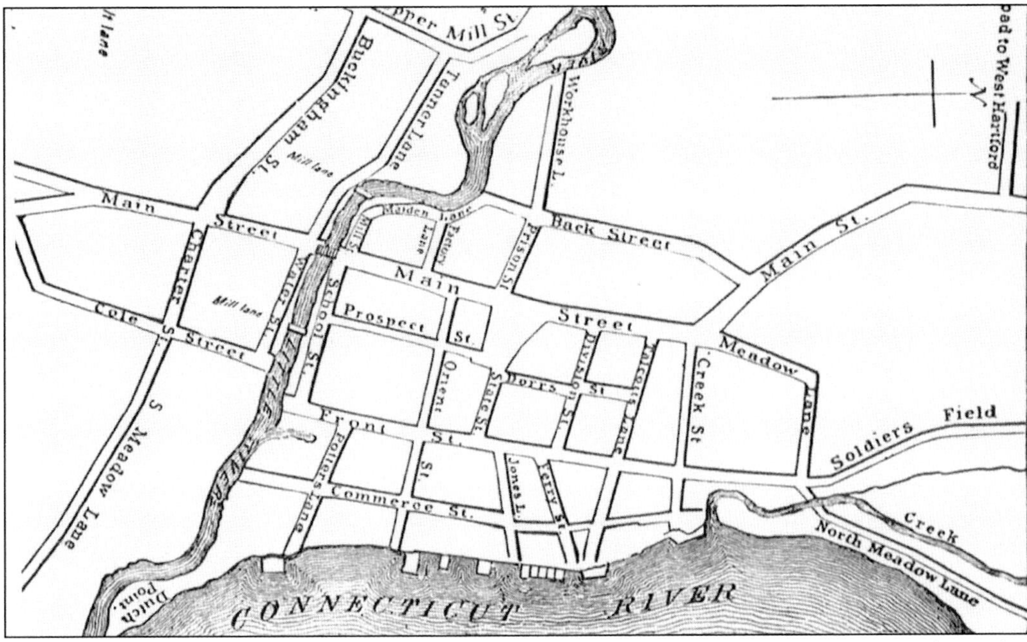

Seen here is a map of the city of Hartford in 1810. The majority of the population of Hartford lived within the area at that point. (Courtesy of the Connecticut Fire Museum Archives.)

Chiefs of the Hartford Fire Department Volunteer Period 1789-1864

Miles Beach	1789-1805
John Carter	1805-1820
J. Hoadley	1820-1825
J.M. Goodwin	1825-1833
W. Hayden	1833-1843
Allyn Stillman	1843-1846
R.G. Drake	1846-1848
Charles Benton	1848-1850
Erastus Hubbard	1850-1852
John Carter	1852-1854
John G. Parsons	1854-1856
Joseph Pratt	1856-1858
Samuel H. Havens	1858-1860
Edward Norton	1860-1862
Horace Billings	1862-1864
Jared J. Butler	1864

This shows all the chiefs of the Hartford Volunteer Fire Department. (Courtesy of the Connecticut Fire Museum Archives.)

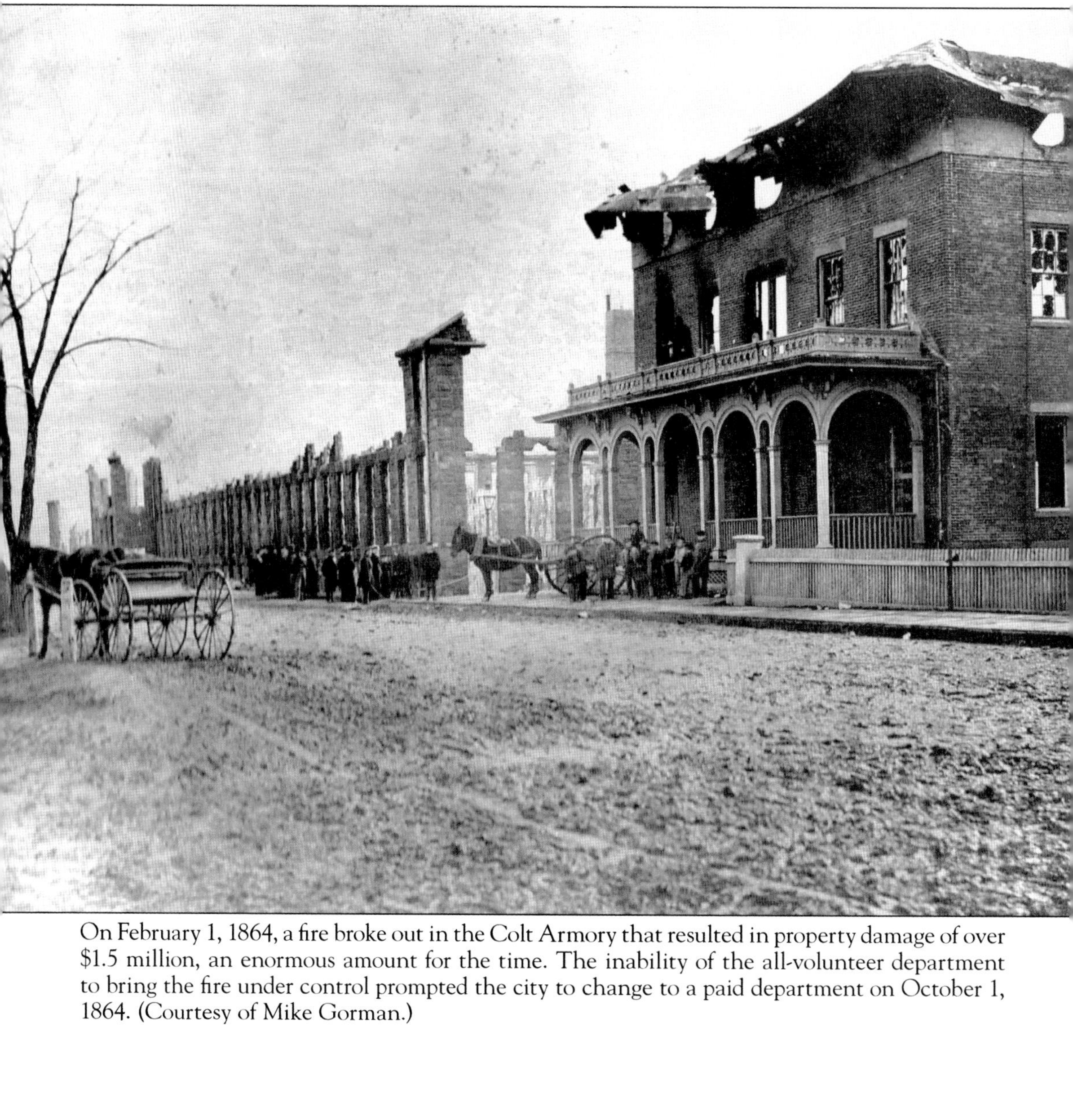

On February 1, 1864, a fire broke out in the Colt Armory that resulted in property damage of over $1.5 million, an enormous amount for the time. The inability of the all-volunteer department to bring the fire under control prompted the city to change to a paid department on October 1, 1864. (Courtesy of Mike Gorman.)

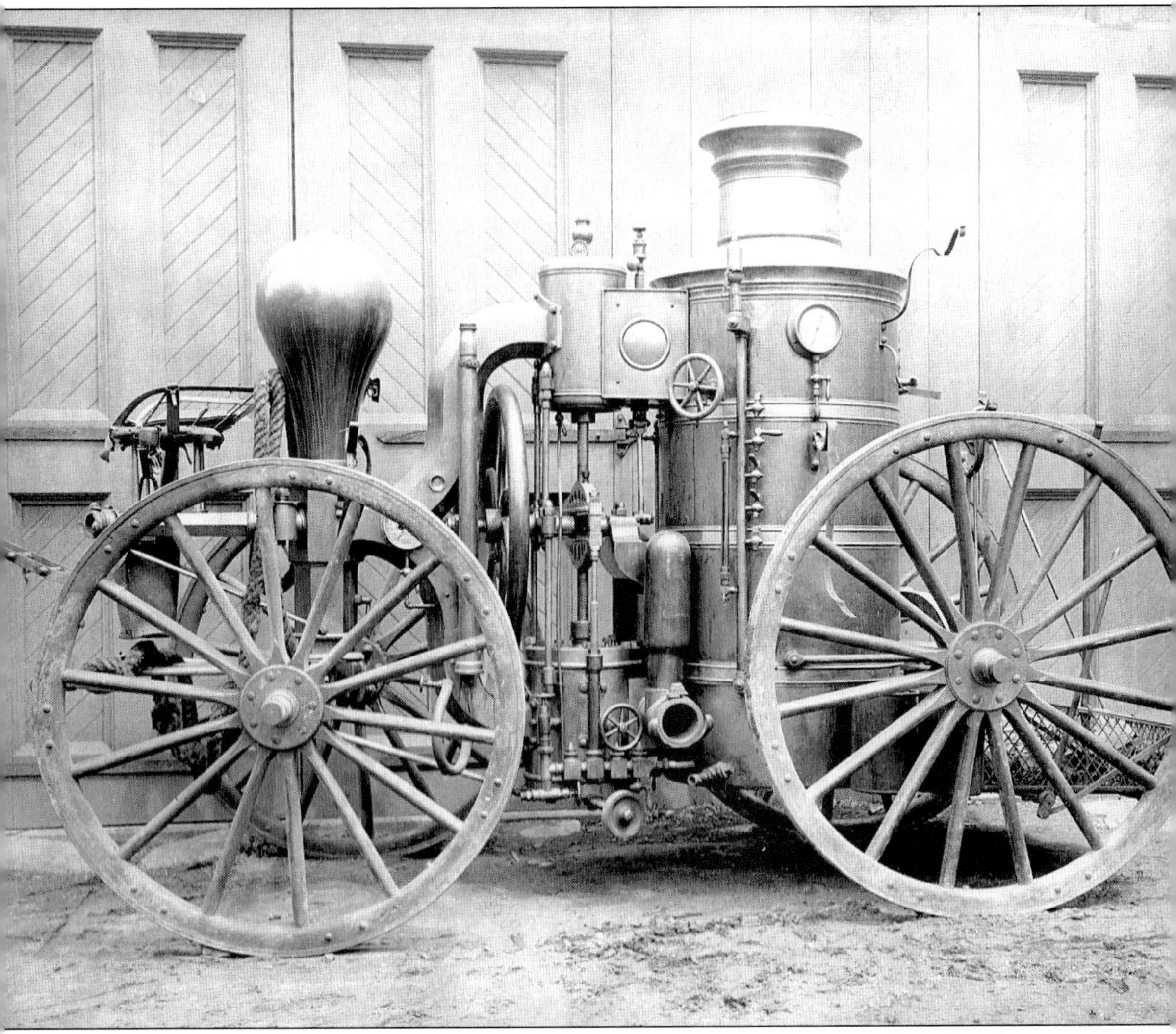

Hartford's first steam pumper, Phoenix No. 3, was purchased in 1861. By 1864, all the hand engines had been retired. The effectiveness of hand engines was severely limited by the number of men needed to work the handles before dropping from exhaustion. (Courtesy of Mike Gorman.)

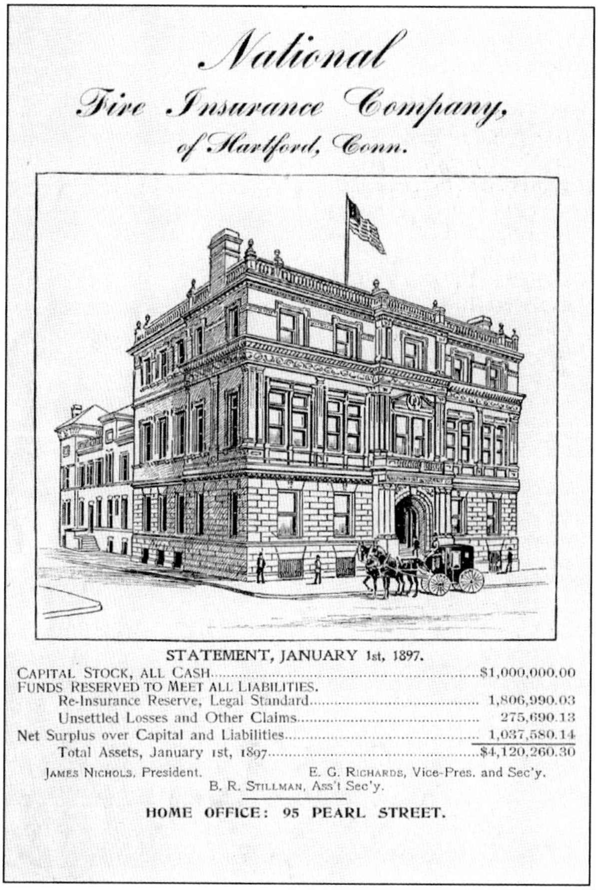

This *c.* 1890 advertisement for one of the fire insurance companies that were headquartered in Hartford includes a statement of the company's financial status. In the volunteer days, fire insurance companies paid a bonus to the first company to arrive at the fire scene. (Courtesy of the Connecticut Fire Museum Archives.)

1860 Organization of the Hartford Fire Department

Engine Companies

Company # 1	**"Charter Oak"**	**37 Main Street**
Company #2	**"Neptune"**	**798 Main Street**
Company #3	**"Phoenix"**	**66 ½ Front Street**
Company #4	**"Damper"**	**126 Main Street**
Company #5	**"Annihilation"**	**3 Church Street**

Hose Companies

Hose Company #1 "Aetna"	**126 Main Street**
Hose Company #2 "Stillman"	**17 Temple Street**

Hook & Ladder Company

Hook & Ladder Company #1	**12 Pearl Street**

Sack and Bucket Company

Company #1 "Hylas"	**12 Pearl Street**

This displays the organization of the Hartford Fire Department in 1860. (Courtesy of the Connecticut Fire Museum Archives.)

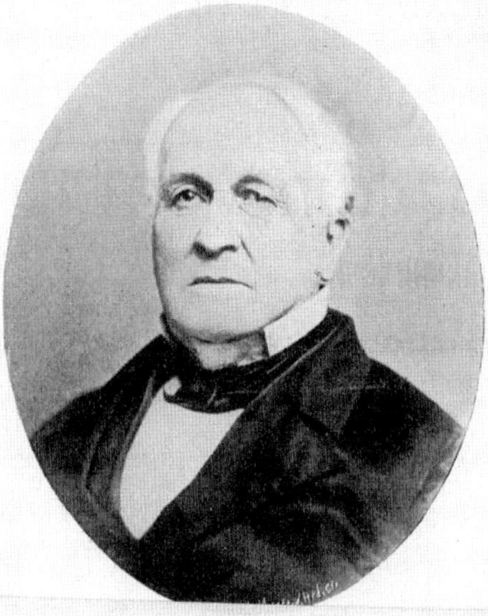

JAMES GOODWIN, Ex-Chief,
Served from 1826 to 1833.

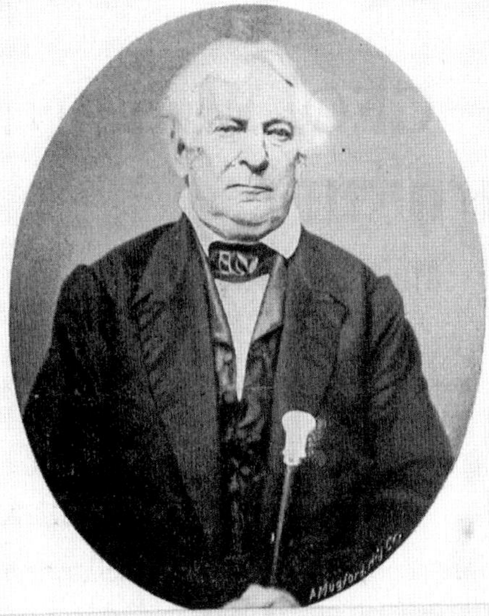

WILLIAM HAYDEN, Ex-Chief,
Served from 1833 to 1843.

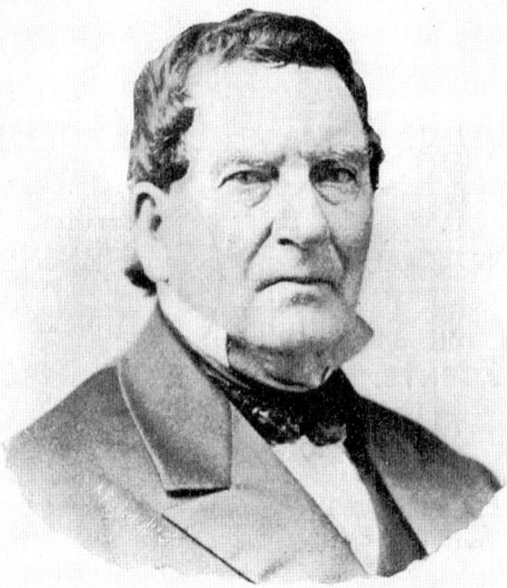

ALLYN S. STILLMAN, Ex-Chief,
Served from 1843 to 1846.

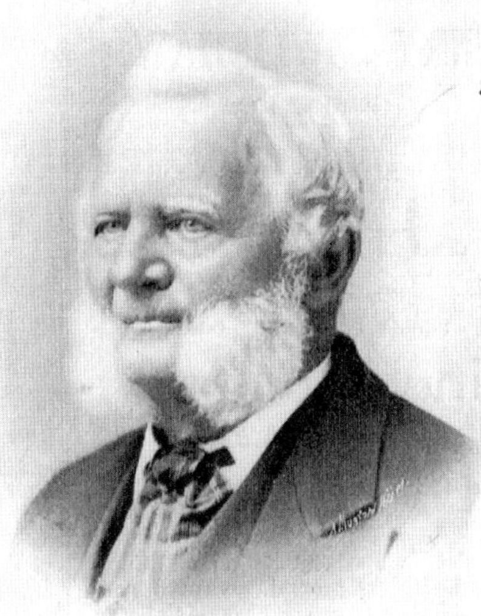

CHARLES BENTON, Ex-Chief,
Served from 1848 to 1850.

Shown here are some of the chiefs of the volunteer fire department in Hartford. (Courtesy of the Alarm Room News.)

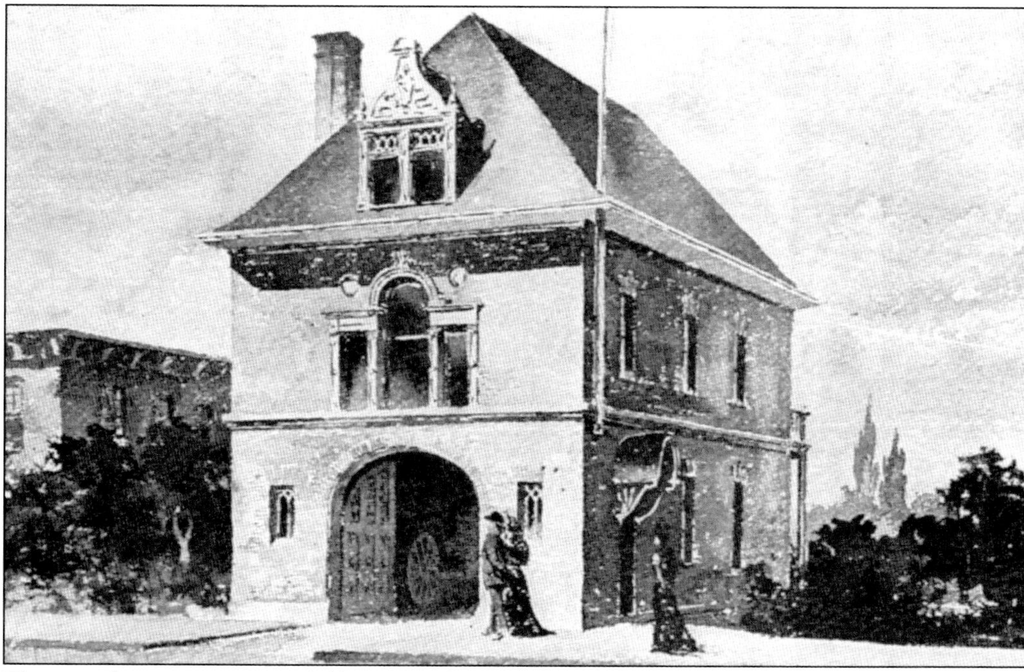

This was the new veterans' clubhouse at the corner of Main and Arch Streets. This building has since been torn down, and the organization is no longer in existence. The Veteran Volunteer Firemen's Association was organized November 25, 1889, and was intended to be made up entirely of ex-members of the old hand engine companies. However, by 1893, the association rules were amended to allow any firefighter with five years of service or more to join. (Courtesy of the Alarm Room News.)

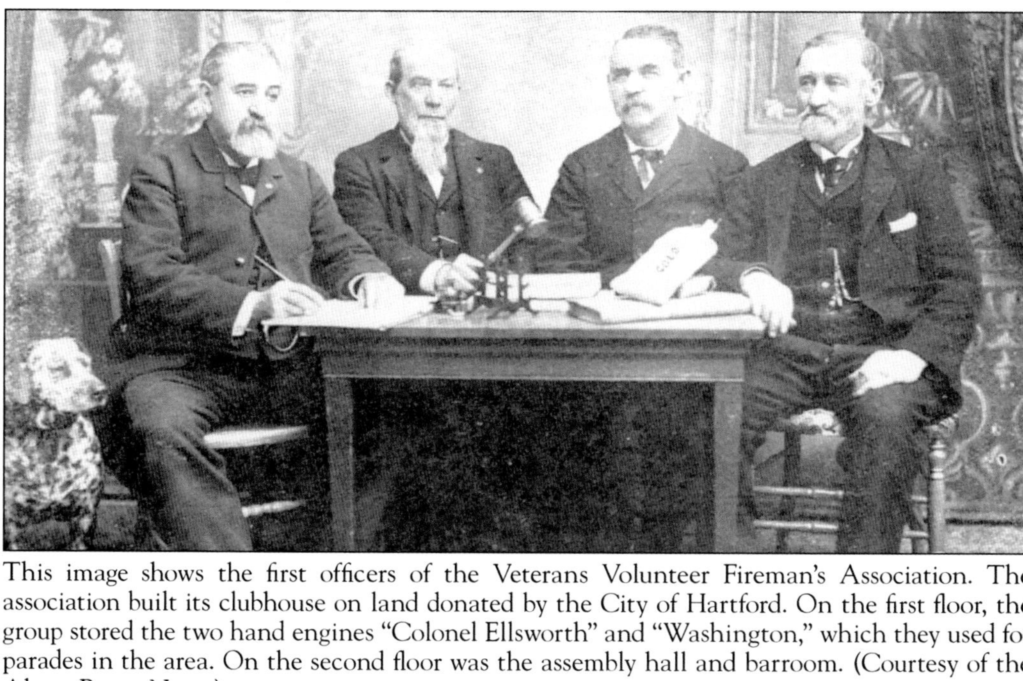

This image shows the first officers of the Veterans Volunteer Fireman's Association. The association built its clubhouse on land donated by the City of Hartford. On the first floor, the group stored the two hand engines "Colonel Ellsworth" and "Washington," which they used for parades in the area. On the second floor was the assembly hall and barroom. (Courtesy of the Alarm Room News.)

Two

THE HORSE-DRAWN ERA

On December 1, 1864, the Hartford Fire Department consisted of a chief engineer, three assistant engineers, a fire marshal, four steam engine companies, two hose companies, and one hook and ladder company. Each steam engine company consisted of a foreman, an engineer, a driver, and nine extra men. Each hose company consisted of a foreman and 12 extra men. The hook and ladder company consisted of a foreman and 20 extra men. The Board of Fire Commissioners set salaries for the employees. They set annual salaries as follows: chief engineer $800, assistant engineer $100, company foreman $50, engineer of each company $700, firemen $60, drivers $450, and extra men: $40. The chief engineer, engineers, and drivers were on duty continuously; all the other members were callmen, essentially employees who were called in only when there was a fire. This actually created a saving in the cost of fighting fires. In the last year of Hartford's volunteer department, 1864, the expenses were $22,456.27 compared to $16,581.12 for 1867, a savings of $5,875.15, or over 26 percent! A large fire bell was purchased in 1867, and this bell summonsed the callmen for three decades. Henry P. Seymour was appointed first chief engineer of the paid department. Henry J. Eaton succeeded him in 1868. Chief Eaton led the department for 35 years, the longest of any Hartford fire chief to date, until his retirement in 1903.

Under Chief Eaton's leadership, the department grew, and many companies were added. In 1872, the Lawrence Steam Engine Company became Hartford's Engine Company No. 5, and Colt's Steam Engine Company, operating out of the Colt's factory complex, was reorganized as Colt's Steam Engine Company No. 6. The first self-propelled steam fire engine was purchased on March 1, 1876, from Amoskeag of Manchester, New Hampshire, and put in service at Engine Company No. 7. This engine, referred to as "Black Seven," cost $5,000, weighed five tons, and could pump 700 gallons per minute. The engine served the city for 37 years, until 1913.

In August 1889, the city purchased the world-famous "Jumbo" from Amoskeag. This extra double size engine was largest engine of its class in the world. Although they were somewhat successful, only 22 "giants" were built. They could reach a top speed of only 10 miles per hour. Its first driver died when he fell off the engine while making a turn, falling under its wheels, and was crushed to death. Another accident occurred resulting in the death of a horse and rider that stumbled and was run over by the lumbering machine.

By 1898, the department consisted of eight steam-powered fire engines, Engines No. 3 and No. 4 being self propelled, and horse-drawn hose wagons. The department also had a chemical engine, hook and ladder aerial truck, and a ladder truck with ground ladders only. All the engine houses were equipped with spring opening doors and automatic harness devices that hitched the horses to the engines or wagons. Their average response time was under two minutes from receipt of the alarm to the apparatus leaving the station. Following Boston's lead in 1852, Hartford was the first city in Connecticut to adopt the Gamewell automatic fire alarm in 1868. Originally boxes were locked with a key. In the event of a fire, the key had to be obtained from a "responsible person living near by" who had to unlock the box and turn in the alarm. Boxes that allowed the public to turn in the alarm without the key were introduced around 1905. By 1910, Hartford had a system of 224 boxes located at all major intersections and important buildings in the city.

By the late 1880s, Hartford had one of the best water systems in the United States. This system supplied water for both households and fire hydrants throughout the city. In 1890, the city had over 500 hydrants. During the late 1890s, the department continued to grow;

Engine Company No. 8 was opened at Park and Affleck Streets in 1896. In 1898, Engine Company No. 7 relocated to 480 Windsor Avenue. In 1900, Chemical Engine Company No. 9 was organized and located at 43 Pearl Street, with Chemical Engine No. 10 at 94 Bond Street and Chemical Engine No. 11 at 3 Sisson Avenue.

The most notable fire of the decade was on May 17, 1895. The bridge spanning the Connecticut River between East Hartford and Hartford burned. Engine Company No. 3's hose wagon and its team were lost in the fire with its two firemen barley escaping.

The beginning of the 20th century saw Hartford's population grow to over 40,000 residents. It was a center of manufacturing in the Northeast, with products known worldwide. Colt firearms, the Gatling gun, Columbia bicycles, Pope-Hartford automobiles, Pratt and Whitney machine tools, and Capewell horseshoe nails were among those products. An extensive streetcar system radiated out of the city to East Hartford, Glastonbury, Manchester, and Unionville. The New York, New Haven and Hartford Railroad and the Central New England Railroad both had facilities in the city and linked Hartford to the rest of the country.

This railroad connection allowed Hartford to send men and equipment from Hartford to Waterbury on a special train on February 2, 1902, when a major fire broke out in Waterbury.

On November 27, 1903, Chief Eaton, fire chief since 1868, retired, and Louis Krug was appointed fire chief. Engine Company No. 12 was placed in service at 243 Smith Street (later renamed South Whitney Street in 1915). The company entered service with a new Amoskeag steam pumper and an American LaFrance chemical and hose wagon. In 1906, $3,000 was appropriated for the purchase of land on Albany Avenue to build a firehouse to cover the western part of the city. On April 1, 1907, Engine Company No. 14 opened at 25 Blue Hills Avenue. The year 1907 saw the arrival of Hartford's first motorized fire apparatus powered by a gasoline engine, a Knox chemical and hose wagon assigned to Engine Company No. 2. In 1908, John C. Moran, formally the engineer of Engine Company No. 4, was promoted to deputy chief. He would become the chief in 1913. Also in 1908, $24,000 was appropriated to construct and equip a fire station, Engine Company No. 15, in the southwest section of the city, at the corner of New Britain and Fairfield Avenues. It opened on April 16, 1909, equipped with an American LaFrance chemical and hose wagon. In 1912, a new Clapp and Jones steam pumper with an American LaFrance gasoline electric tractor, costing $9,000, was ordered for Engine Company No. 1. Engine Company No. 3 moved from its old quarters on Front Street to a new three-bay station at 90 Market Street. Also in 1912, Truck Company No. 2 was transferred from 275 Pearl Street to Engine Company No. 15's house on New Britain Avenue.

The self-propelled "Jumbo," purchased in 1889, was rebuilt in 1915 and converted to a tractor-drawn machine by removing its self-propelling machinery and installing an American and British gasoline electric tractor that used a gasoline engine to drive a generator powering an electric motor on each front wheel, very similar to trolley cars of the day. With the arrival of more motorized fire apparatus, the department was fast approaching conversion to a totally motorized department.

Henry P. Seymour was the first chief of Hartford's new paid department, serving from 1864 to 1868. Henry J. Eaton succeeded him and served as chief for 35 years. (Courtesy of the Alarm Room News.)

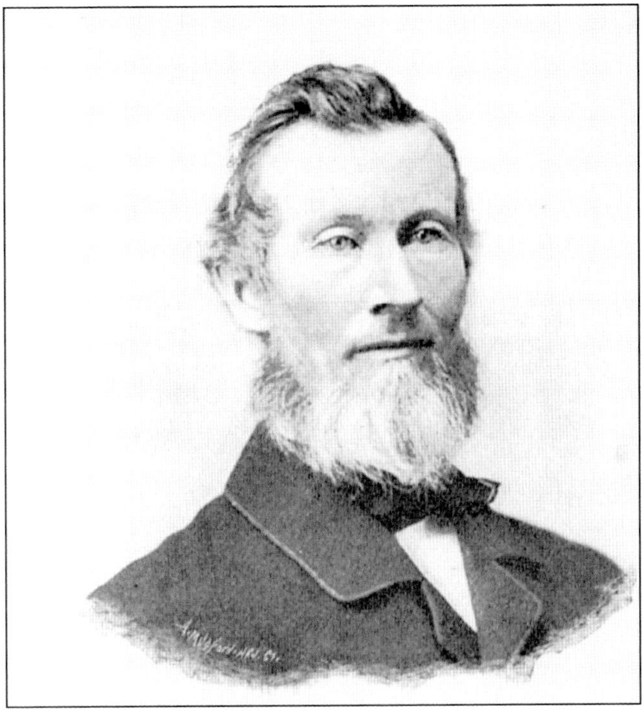

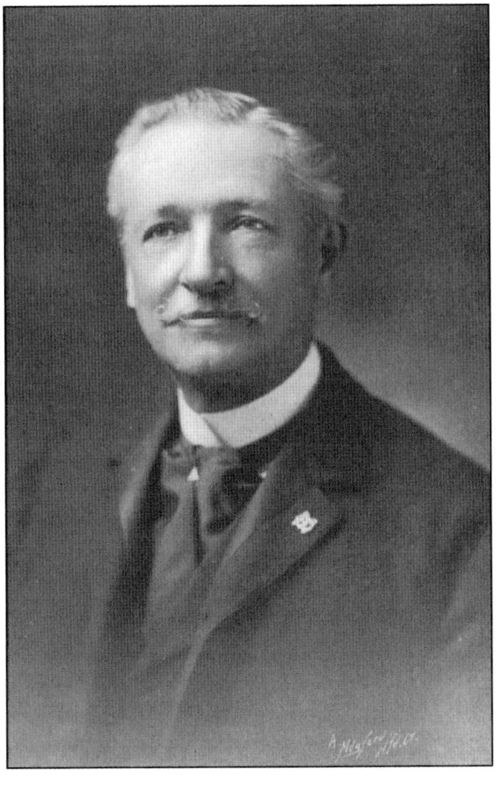

Charles L. Billings, president of the Board of Fire Commissioners in 1898, was a native of Vermont. Billings began his career working at the Colt Firearms factory as a tool and die maker. In 1869, he established the well-known firm of Billings and Spencer that made hand tools. (Courtesy of the Alarm Room News.)

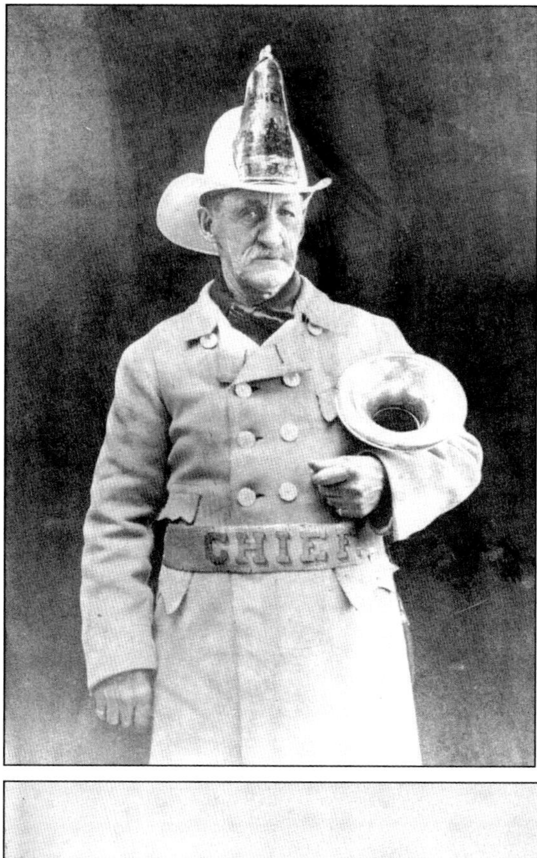

Henry J. Eaton began his career with the fire department on April 5, 1851, as a member of Protection Engine Company No. 1. When the department was changed from volunteer to paid service, he was appointed first assistant engineer under Chief Henry P. Seymour. (Courtesy of Mike Gorman.)

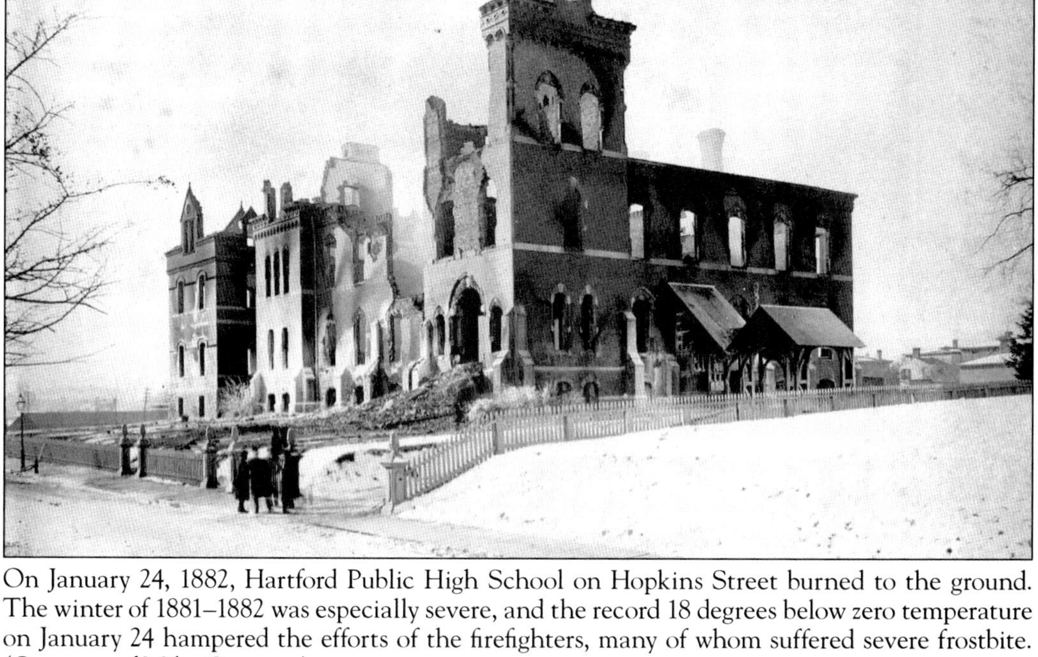

On January 24, 1882, Hartford Public High School on Hopkins Street burned to the ground. The winter of 1881–1882 was especially severe, and the record 18 degrees below zero temperature on January 24 hampered the efforts of the firefighters, many of whom suffered severe frostbite. (Courtesy of Mike Gorman.)

On July 7, 1884, fire was discovered in the belfry of the South Congregational Church on Main Street, causing $17,000 worth of damage to the building. This was not the first fire in that building, nor was it the last. Evidence of the fire can still be seen today in the belfry. (Courtesy of Mike Gorman.)

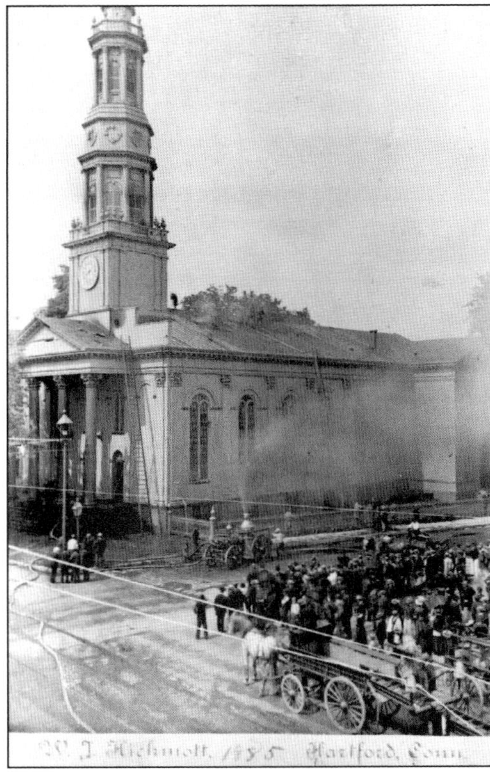

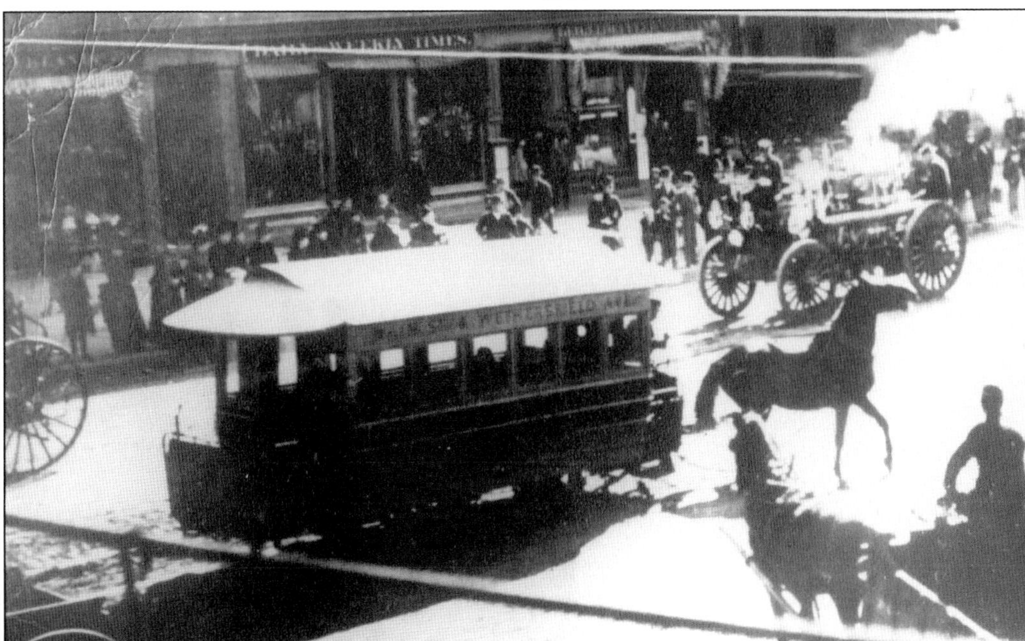

Engine Company No. 3's Amoskeag self-propeller "Jumbo" is shown in this 1889 view with a Hartford and Wethersfield Horse Railroad horsecar on Main Street at Grove Street. This trial run, on October 18, 1889, created quite a stir on Main Street. (Courtesy of the F. S. Bennett Collection.)

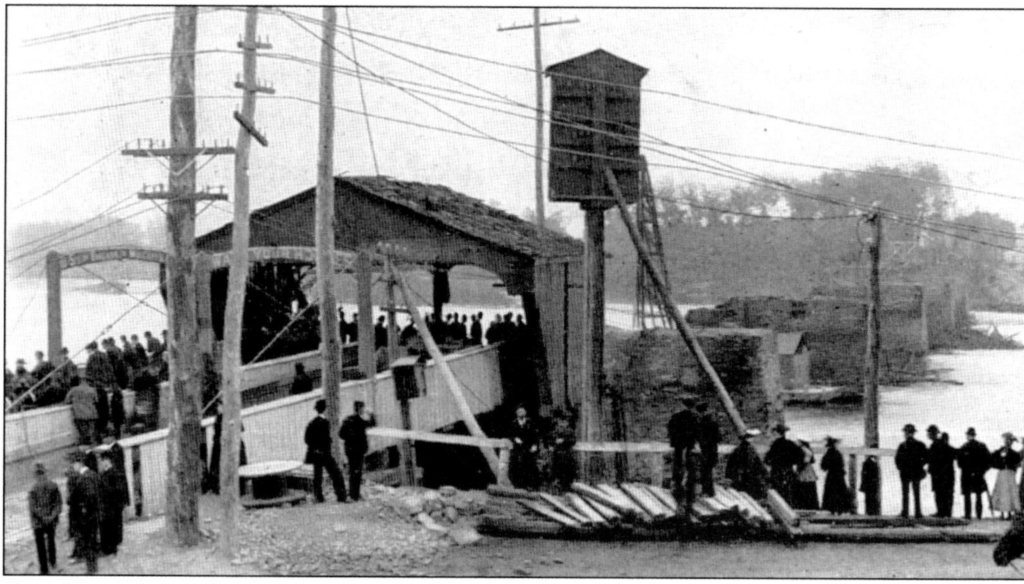

On May 17, 1895, the wooden bridge spanning the Connecticut River, between East Hartford and Hartford, burned in a spectacular fire. The fire started around 7:00 p.m. on the east side of the river. A still alarm called out Engine No. 3's hose wagon, which drove to the East Hartford end of the bridge where the fire first appeared. About 7:15 p.m., box 29, at Morgan and Front Streets, was struck for the fire, calling out the entire department of seven engines and the ladder truck company. The fire burned rapidly through the 75-year-old pine timber. The hose wagon and horses of Engine Company No. 3 were lost inside the burning bridge, and the driver and firemen barely reached safety. More than 20,000 people were reported to have witnessed the fire, as reported in the *Hartford Courant*. (Courtesy of the Connecticut Fire Museum Archives.)

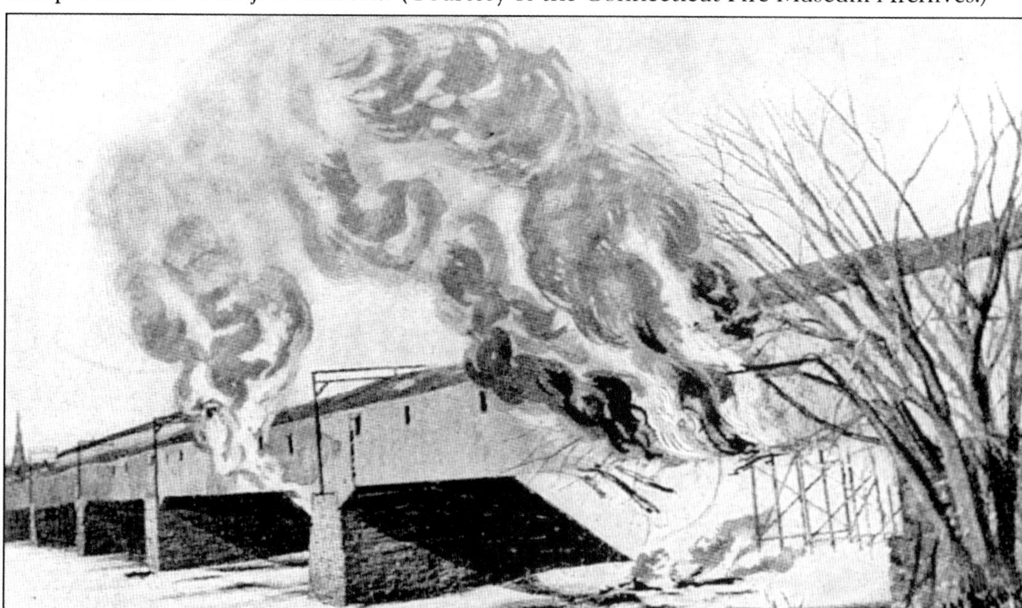

Artists also drew pictures of the bridge fire, like this rendition as seen from the East Hartford side of the river. (Courtesy of the Connecticut Fire Museum Archives.)

Fire headquarters, located behind the State Savings Bank at 43 Pearl Street, is seen here in 1898, with Chemical Engine Company No. 1 and Chief Henry J. Eaton, to the right, in his buggy. Chemical No. 1 was disbanded in 1900 and reorganized as Chemical Engine Company No. 9. In 1911, this unit was renamed Squad A, which responded to all box alarms in the city. (Courtesy of the Alarm Room News.)

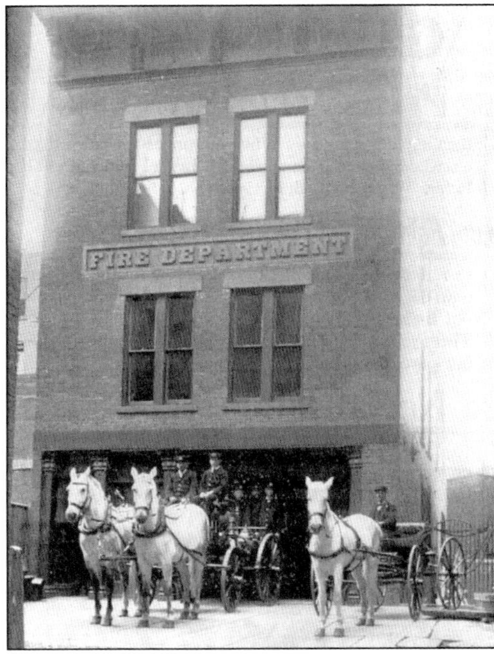

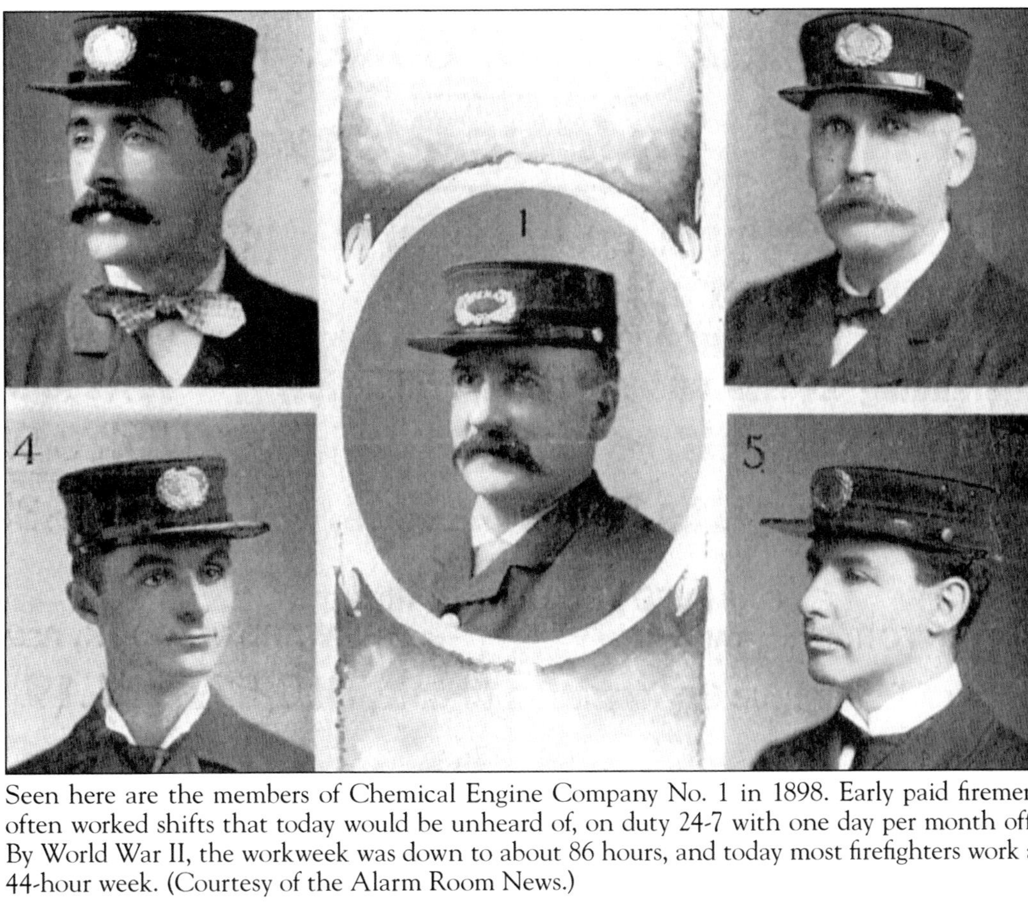

Seen here are the members of Chemical Engine Company No. 1 in 1898. Early paid firemen often worked shifts that today would be unheard of, on duty 24-7 with one day per month off. By World War II, the workweek was down to about 86 hours, and today most firefighters work a 44-hour week. (Courtesy of the Alarm Room News.)

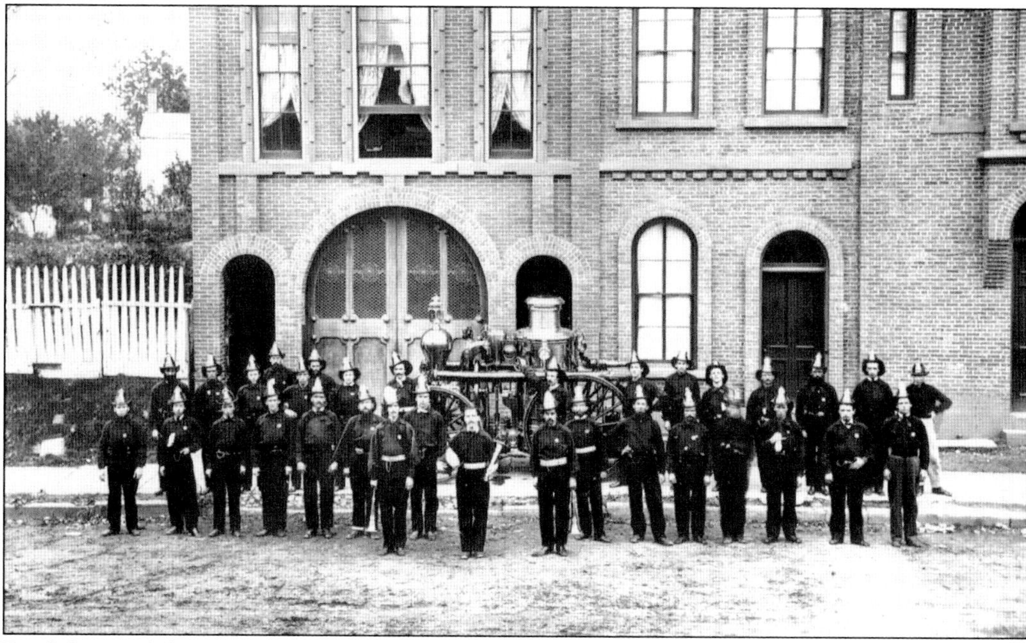

Firefighters of the Charter Oak Steam Engine Company No. 1, located at 55 Main Street, and their apparatus are shown in this 1865 photograph. (Courtesy of Mike Gorman.)

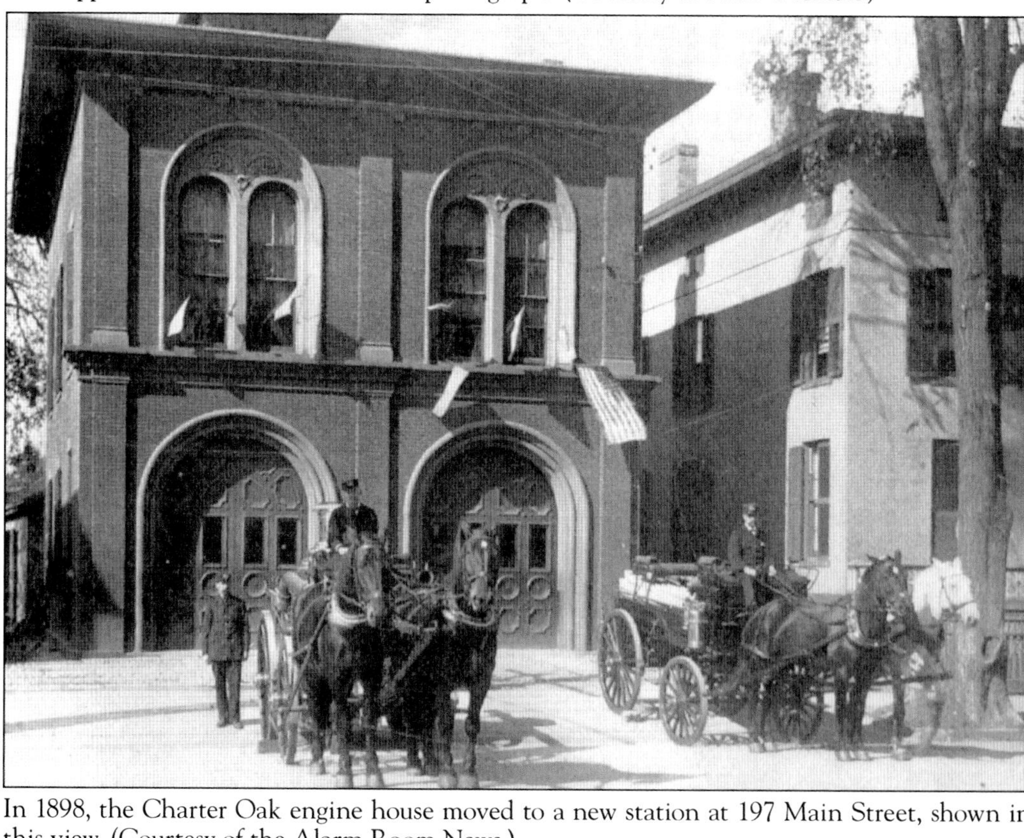

In 1898, the Charter Oak engine house moved to a new station at 197 Main Street, shown in this view. (Courtesy of the Alarm Room News.)

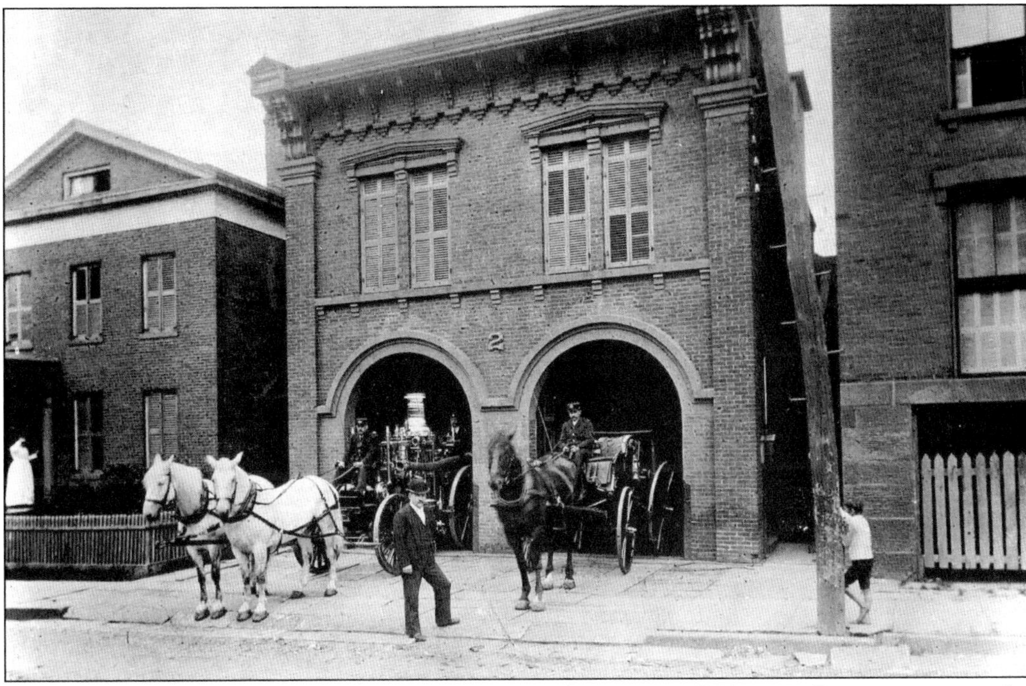

Hope Steam Fire Engine Company No. 2, organized in 1864 at 5 Pleasant Street, is seen in this 1898 view. In 1911, the company relocated to a new three-bay station at 45 Windsor Avenue. (Courtesy of Mike Gorman.)

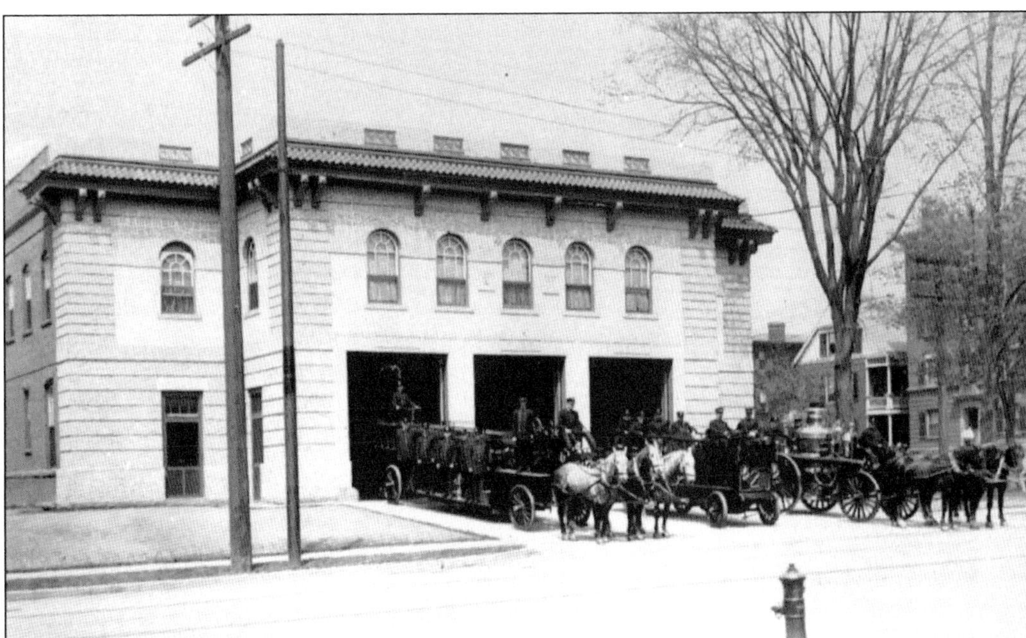

Engine Company No. 2 moved into this new station at 45 Windsor Avenue in 1911. In 1928, Windsor Avenue was renamed Main Street, and the street address became 1515 Main Street. This station remains in service in 2006. (Courtesy of Mike Gorman.)

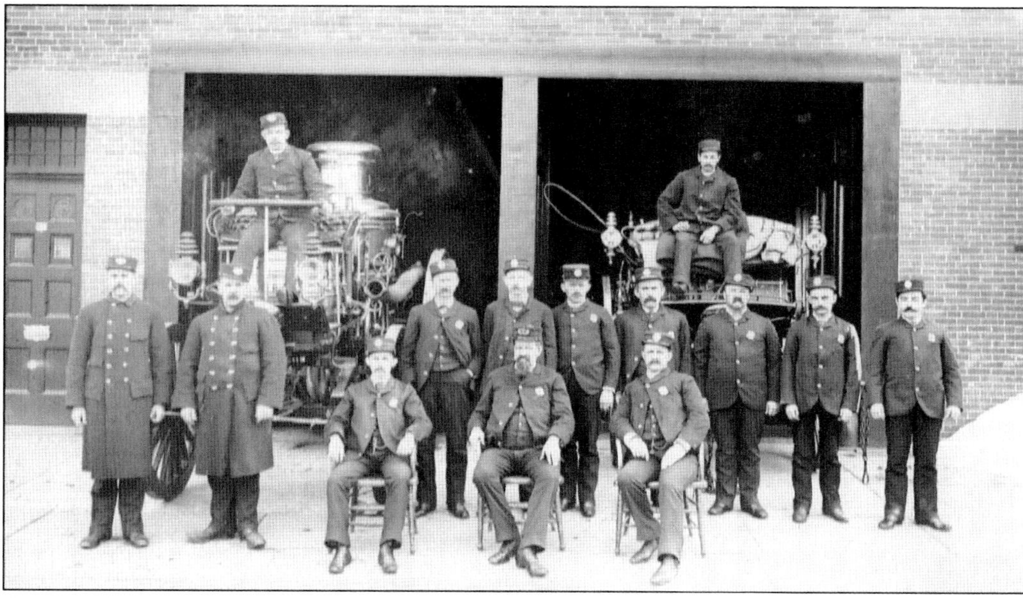

Organized in 1864 as Phoenix Engine, Engine Company No. 3 was located at 124 Front Street. This company was disbanded in 1876 and reorganized as Engine No. 7. As such, they were equipped with the first self-propelled fire engine in the United States, "Black Seven." This engine cost $5,000 and remained in service until 1913. In 1890, Engine No. 7 was disbanded, and Engine No. 3 was reorganized. The year 1912 saw a new station built for Engine No. 3, located at 90 Market Street. (Courtesy of Mike Gorman.)

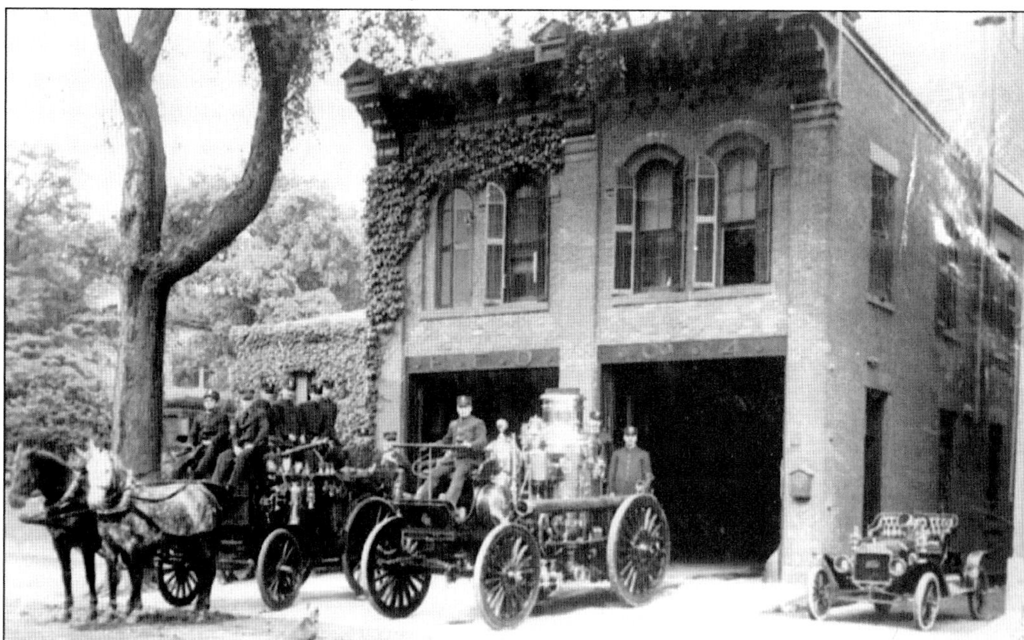

Engine Company No. 4 is shown at its original location at 60 Ann Street, with the "Pride of Hartford," its Amoskeag self-propelled steam pumper. This was the last self-propeller purchased by the city. Engine Company No. 4 moved to the new fire headquarters at 275 Pearl Street in 1927. In 1999, Engine Company No. 4 was disbanded. (Courtesy of Mike Gorman.)

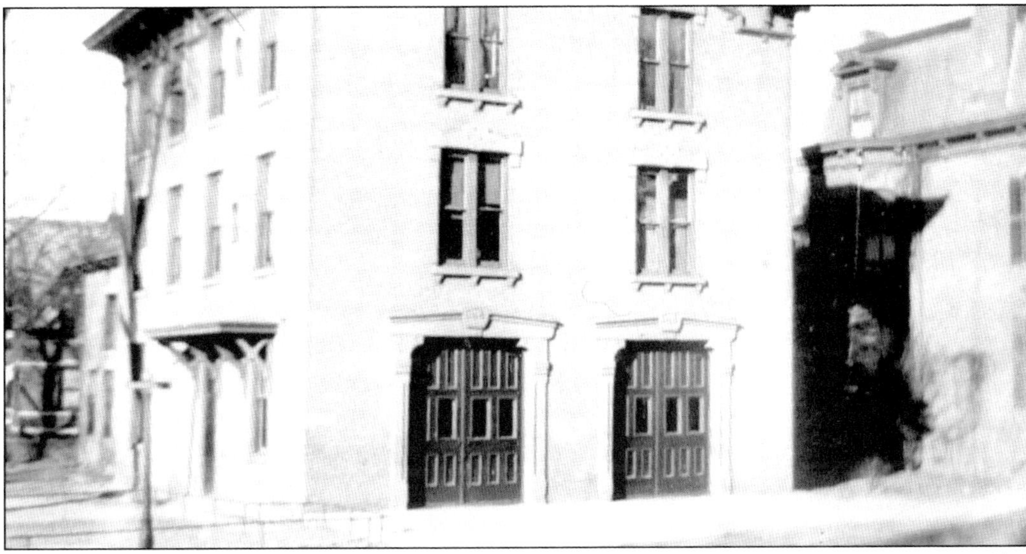

Lawrence Steam Fire Engine Company No. 5 was organized in 1871. Engine Company No. 5 has been located at 129 Sigourney Street since 1871. In 1939, a new station was built on the same site. When an alarm came into this station, a number of events occurred. The stable doors opened automatically, freeing the horses that were trained to walk to their position under the harness. The firefighters climbed on the apparatus, the driver pulled a cord that released the harness suspended on the ceiling, dropping it onto the horses, and then a firemen snapped the harness buckles to complete the operation. In less than two minutes, the steamer and hose wagon rolled out the door. The quick release harness was invented by R. G. Armstrong of Wichita, Kansas, in the 1890s. (Courtesy of the Alarm Room News.)

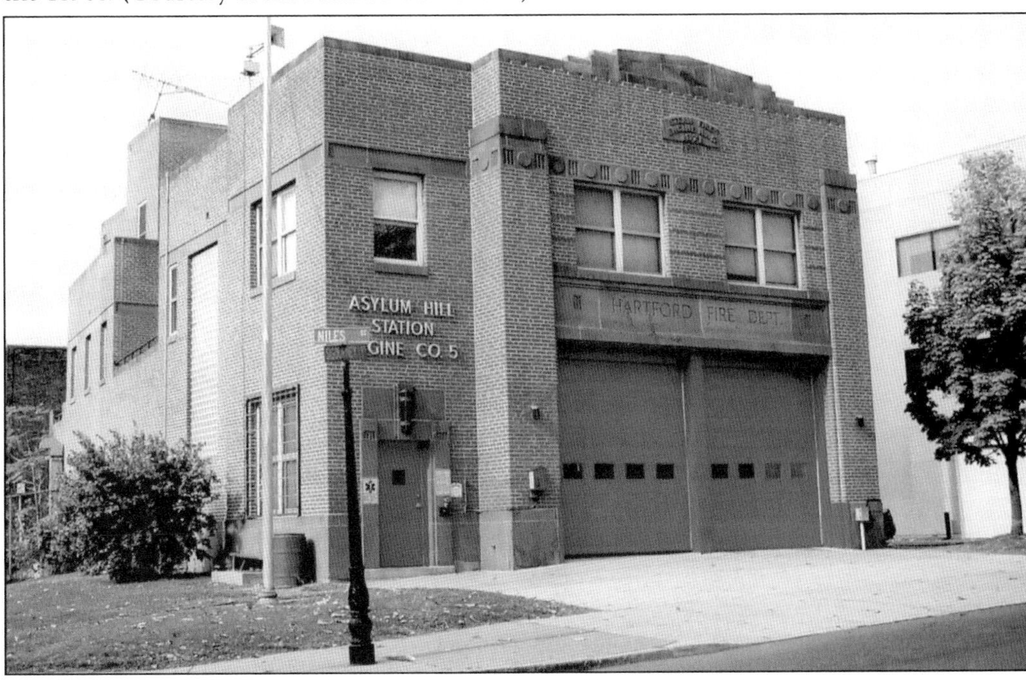

This is a 2006 photograph of Engine Company No. 5's quarters. (Courtesy of the Connecticut Fire Museum Archives.)

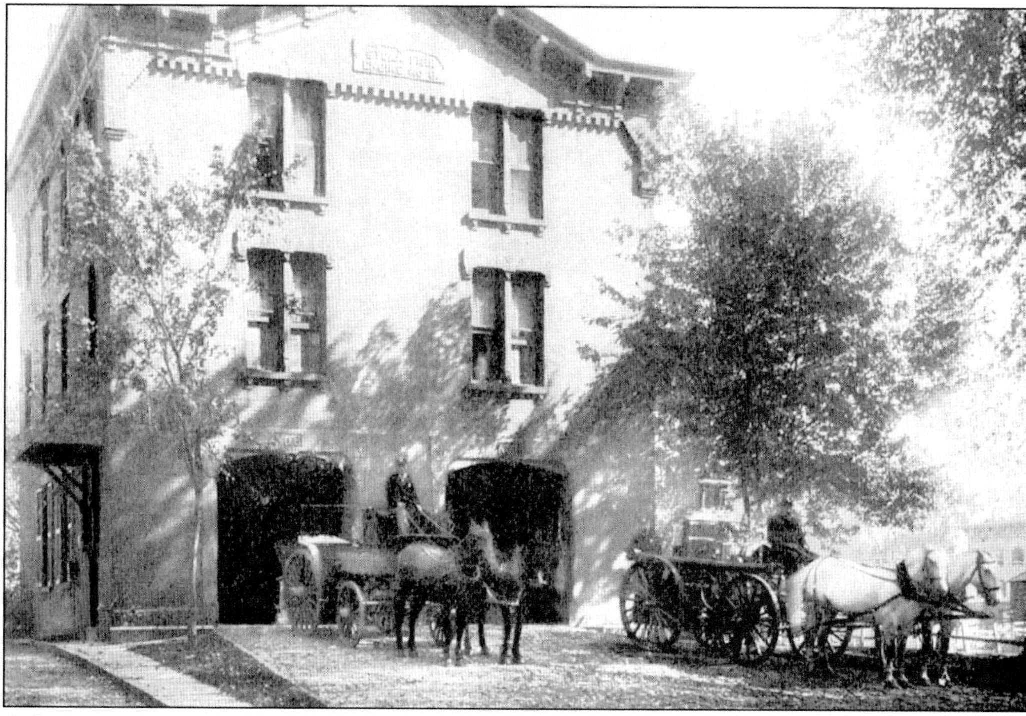

Colt Steam Fire Engine Company No. 6 originally operated in a building at Colt's factory. In 1870, this station at 98 Huyshope Avenue was opened. The company remained there until a new station was opened at 37 Huyshope Avenue in 1915. In 1982, Engine Company No. 6 was disbanded to form Tactical Unit No. 2. (Courtesy of the Alarm Room News.)

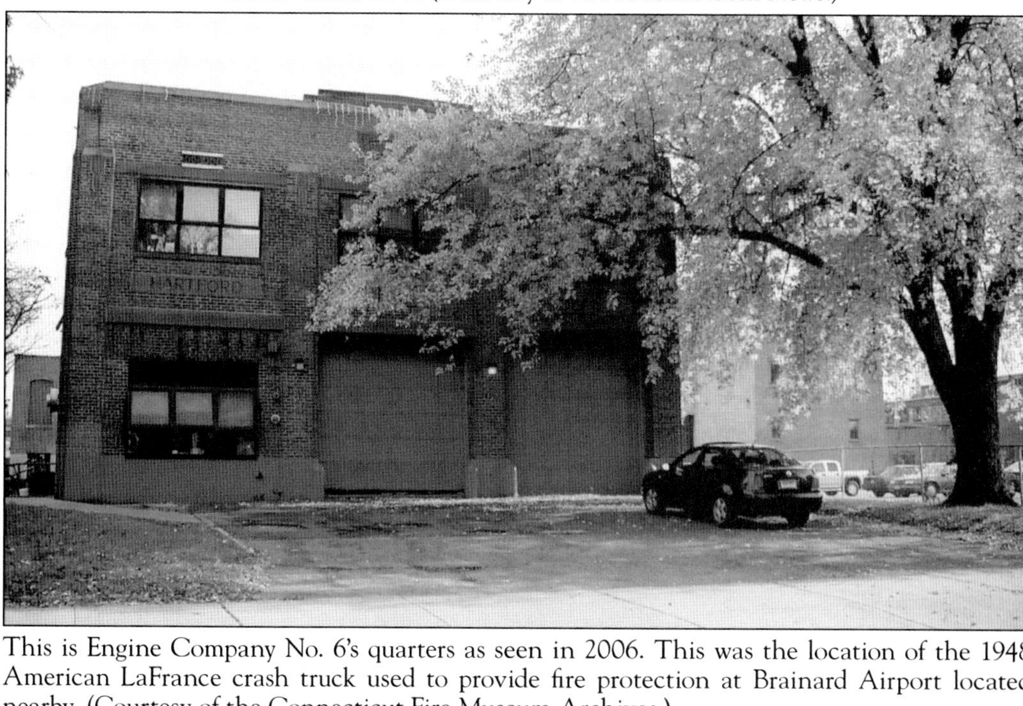

This is Engine Company No. 6's quarters as seen in 2006. This was the location of the 1948 American LaFrance crash truck used to provide fire protection at Brainard Airport located nearby. (Courtesy of the Connecticut Fire Museum Archives.)

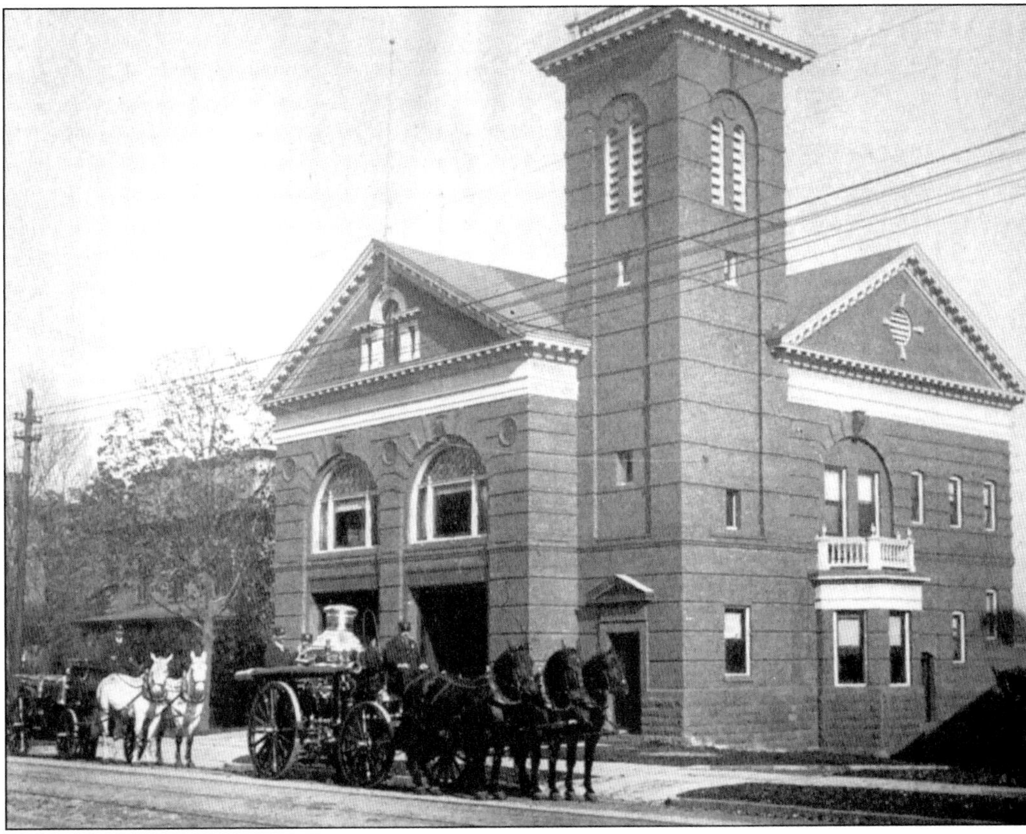

This is Engine Company No. 7's station at 2384 Main Street around 1898. Built in 1895, this station was occupied until 1961 when Engine No. 7 and Engine No. 3 were moved to a new facility at Clark and Westland Streets. (Courtesy of the Alarm Room News.)

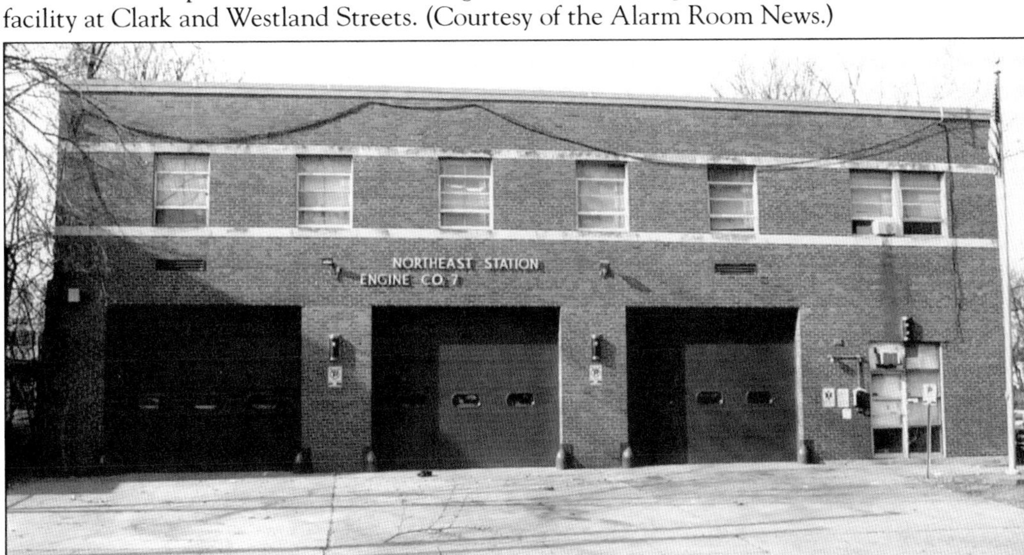

This photograph shows a 2006 view of Engine Company No. 7's quarters. Engine Company No. 3 was disbanded in 1982. (Courtesy of the Alarm Room News.)

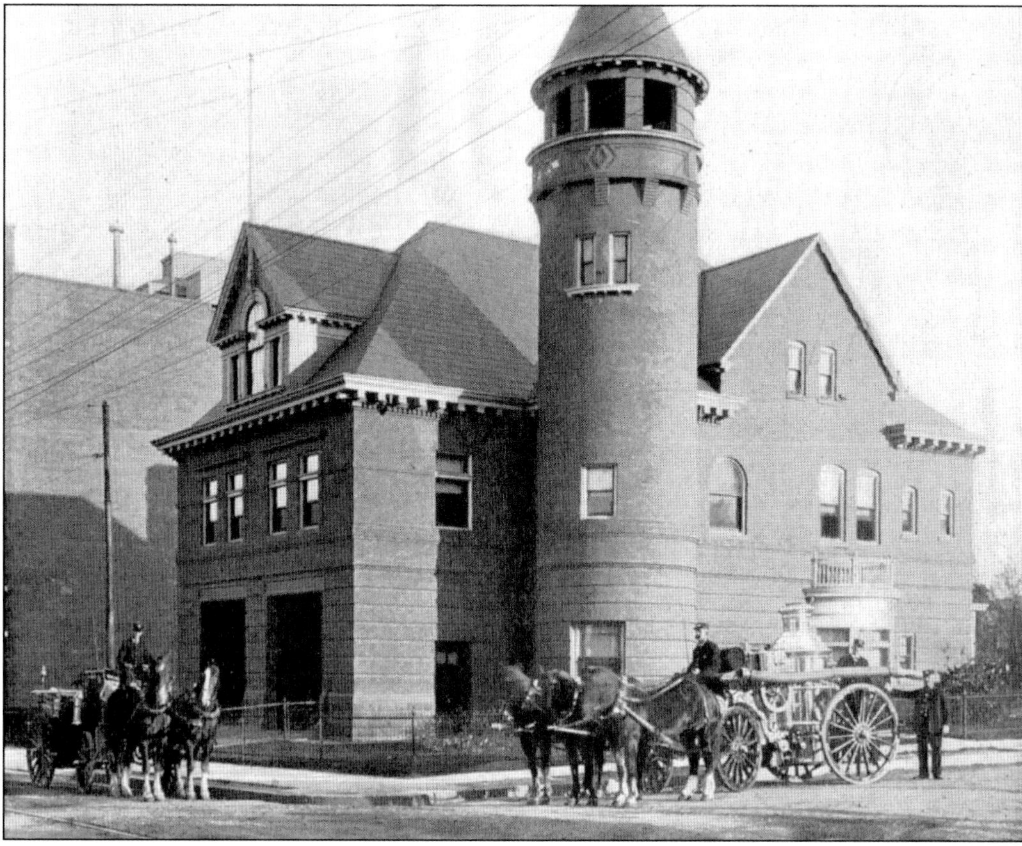

Engine Company No. 8, at 721 Park Street, opened in 1896. The company had a First Class Clapp and Jones engine that pumped at 900 gallons per minute. This building was demolished in 1961, and a new station was built on the same site. (Courtesy of the Connecticut Fire Museum Archives.)

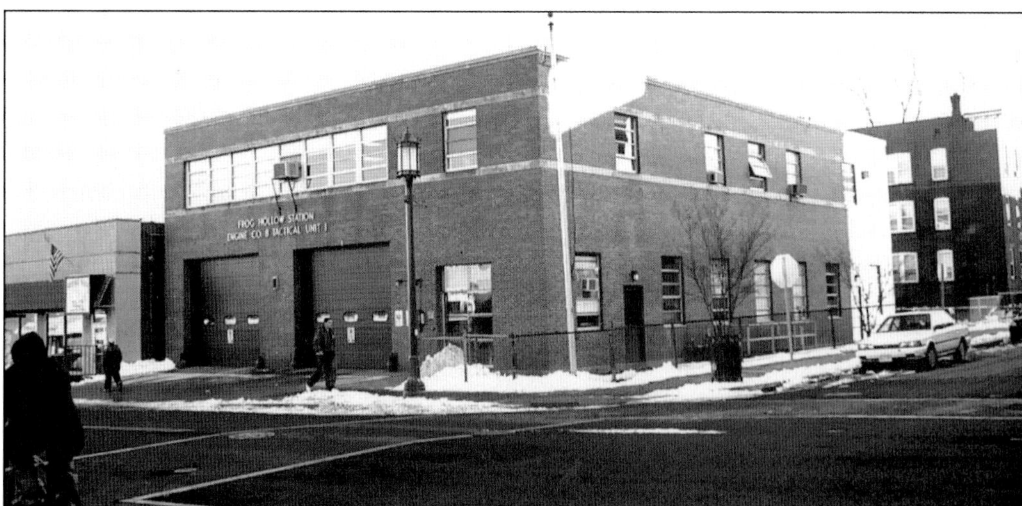

Here one can see a 2006 view of Engine Company No. 8's quarters on Park and Affleck Streets. (Courtesy of the Connecticut Fire Museum Archives.)

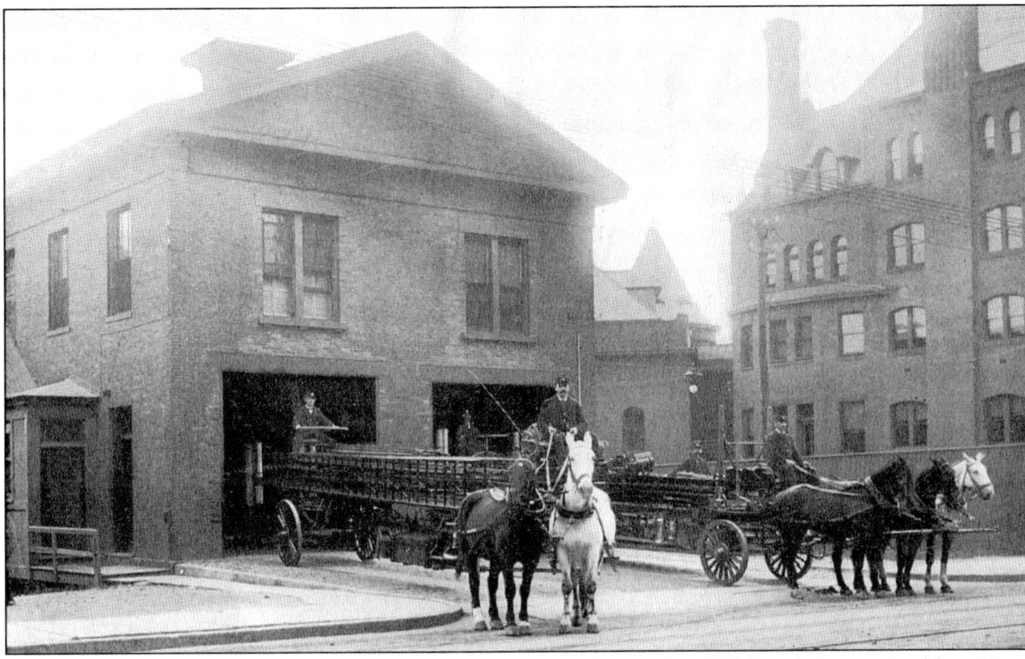

Hayden Hook and Ladder Truck Company No. 1 was located at 275 Pearl Street. Ladders No. 1 and No. 2 are shown in this 1898 view. Ladder No. 2 was relocated to Engine Company No. 15's quarters on Fairfield Avenue in 1912. The second floor housed the families of the driver and tillerman. This is in the day when firefighters were on duty 24-7 with one day per month off. (Courtesy of the Alarm Room News.)

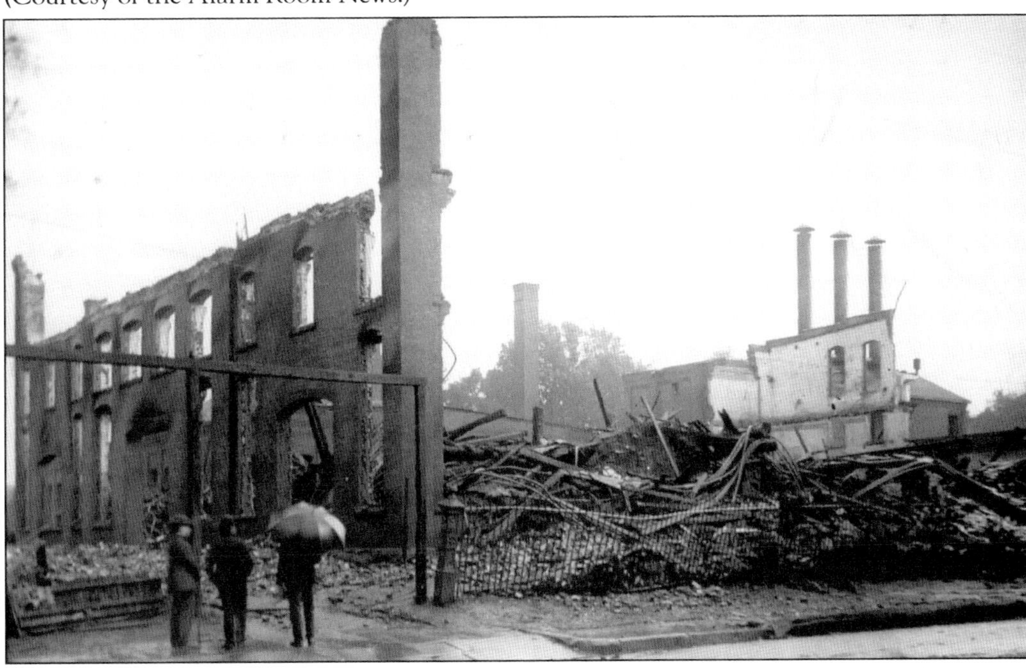

At 2:05 a.m., on July 3, 1902, box 313 was struck for a fire at the Capewell Horse Nail Company on Charter Oak Avenue. The company later rebuilt on the same site and operated until the 1970s, still making horseshoe nails! (Courtesy of Mike Gorman.)

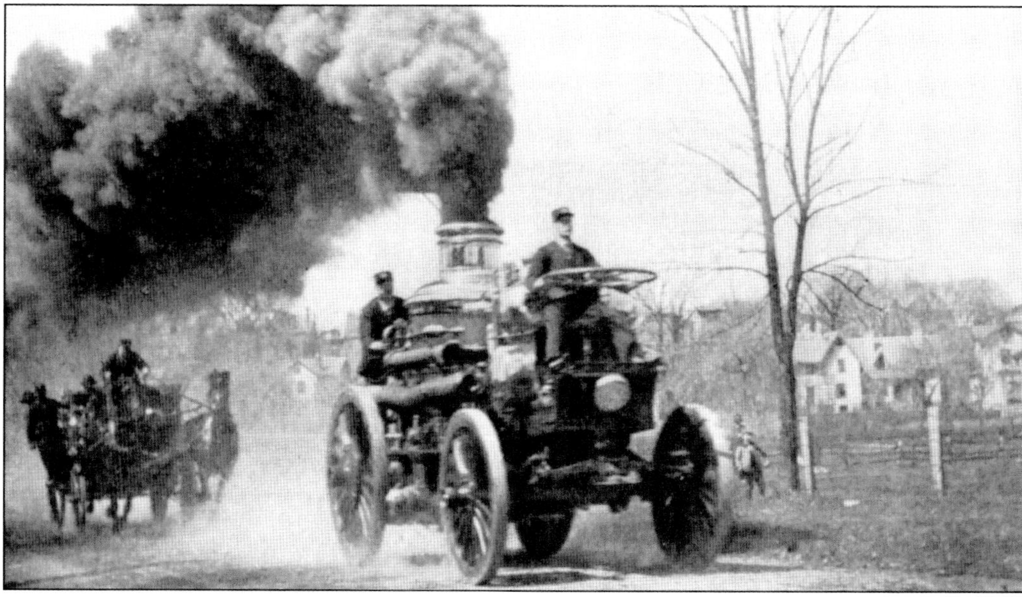

Engine Company No. 3's self-propelled unit "Jumbo" was built by the Manchester Locomotive Works of Manchester, New Hampshire, in 1889. In 1877, Manchester acquired Amoskeag Manufacturing Company, a builder of textile weaving machines and fire engines. Only 22 self-propelled engines were built. They could reach a fantastic top speed, without horses, of 10 miles per hour. This was an enormous machine for its time, weighing over 17,000 pounds, and was rated at 1,350 gallons per minute. In 2006, the age of modern and highly complex fire apparatus, it is important to remember that the old steamers were very effective at pumping water. Only within the past 20 years has modern fire apparatus surpassed the steam pumpers in the amount of water they could supply for a fire. (Courtesy of the Connecticut Fire Museum Archives.)

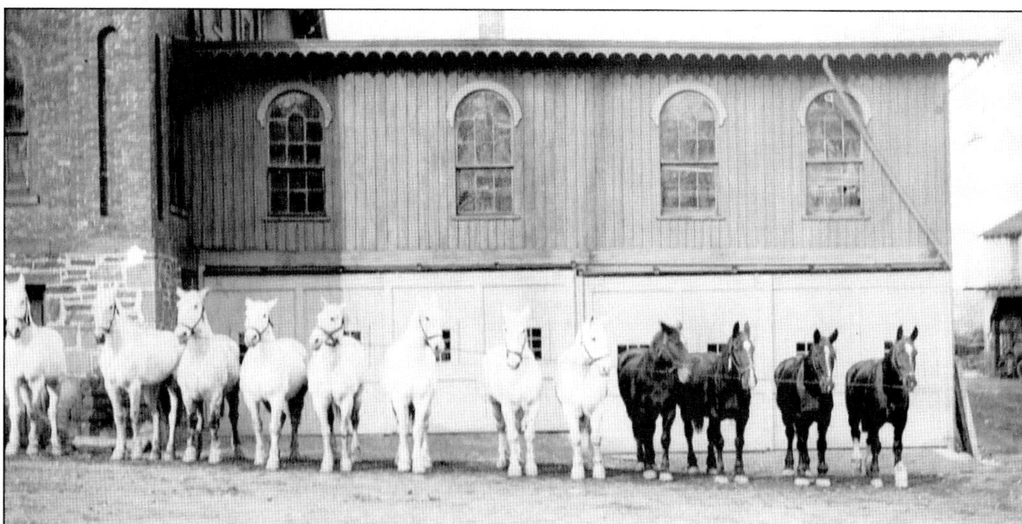

Each horse in the department had a name and was given very good care, munching down 35 pounds of grain and an assortment of horse goodies such as carrots or apples each day. Some of Hartford's soon-to-be-retired fire horses are seen at Colts Park in 1914. In this era, fire stations were basically stables. When horses were finally retired in the 1920s, many of the health hazards associated with stables also disappeared. Firefighters no longer lived in barns. (Courtesy of Mike Gorman.)

1910
Stations and Apparatus

Engine Co. #1	**1900 Amoskeag 550 gpm steam pumper** **Hose Wagon**
197 Main Street	
Engine Co. #2	**1880 Clapp & Jones 1100 gpm steam pumper** **1907 Knox Chemical & Hose Wagon**
5 Pleasant Street	
Engine Co. #3	**1889 Amoskeag 1350 gpm self propeller steam pumper** **Hose Wagon**
124 Front Street	
Engine Co. #4	**1901 Amoskeag 900 gpm self propeller steam pumper** **Hose Wagon**
60 Ann Street	
Engine Co. #5	**1872 Amoskeag 700 gpm steam pumper** **Hose Wagon**
129 Sigourney Street	
Engine Co. #6	**1868 Amoskeag 700 gpm steam pumper** **Hose Wagon**
98 Huyshope Avenue	
Engine Co. #7	**1896 Clapp & Jones 900 gpm steam pumper** **Hose Wagon**
480 Windsor Street	
Engine Co. #8	**1896 Clapp & Jones 900 gpm steam pumper** **Hose Wagon**
721 Park Street	
Chemical Engine Co. #9 **43 Pearl Street**	**1896 Holloway Chemical & Hose Wagon**
Chemical Engine Co. #10 **94 Bond Street**	**1900 Holloway Chemical & Hose Wagon**
Chemical Engine Co. #11 **3 Sisson Avenue**	**1900 Holloway Chemical & Hose Wagon**
Engine Co. #12	**1904 Amoskeag 700 gpm steam pumper** **1904 American-LaFrance Chemical & Hose Wagon**
243 Smith Street **(Street name changed to South Whitney Street in 1915)**	
Engine Co. #14	**1907 Clapp & Jones 700 gpm steam pumper** **Hose Wagon**
25 Blue Hills Avenue	
Chemical Engine Co. 15 **8 Fairfield Avenue**	**1909 American-LaFrance Chemical & Hose Wagon**
Ladder Co. #1	**1899 Gleason & Bailey 75' tiller aerial**
Ladder Co. #2	**1873 Leverich cities service tiller ladder truck**
275 Pearl Street	

This is a list of stations and apparatus of the Hartford Fire Department in 1910. (Courtesy of the Connecticut Fire Museum Archives.)

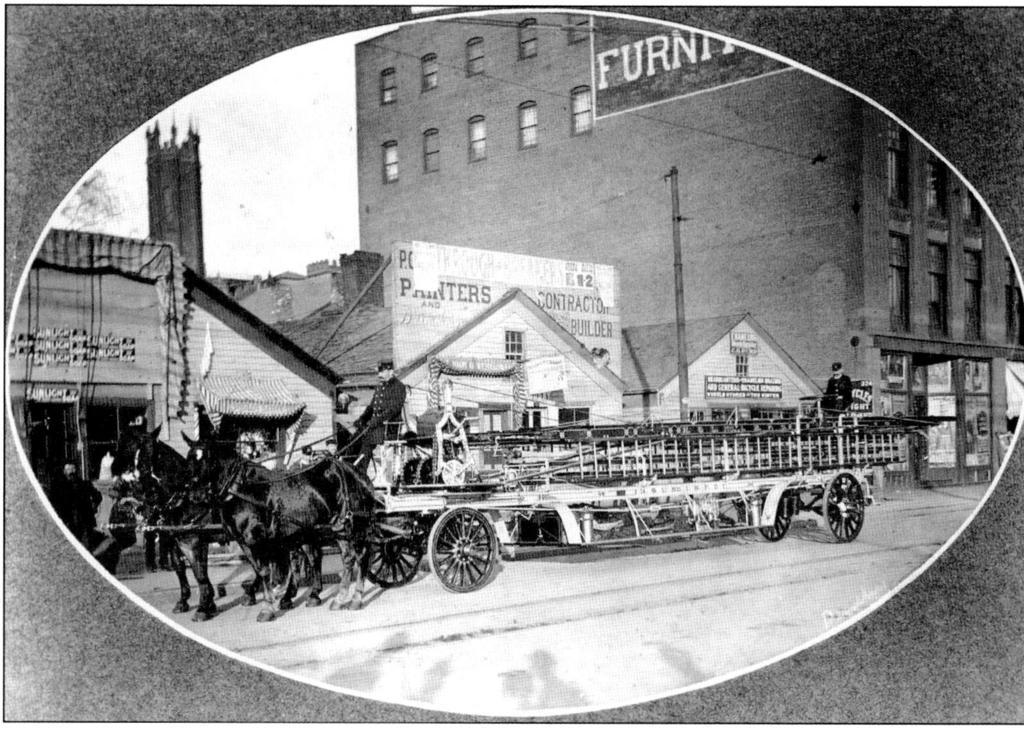

Ladder Company No. 1's driver and tillerman, along with their faithful team, are shown in this 1900 view. Fire horses were not of any particular breed, but they had to be strong and fast. They were usually in matched pairs with only aerial ladders and some of the very large steamers having three-horse hitches. All other equipment, with the exception of the chief's buggy, used two horses. (Courtesy of Mike Gorman.)

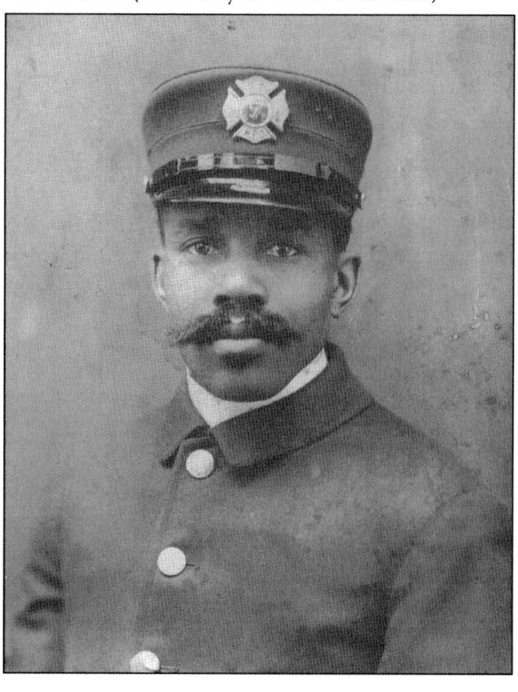

William Henry Jacklyn was the first African American to join the Hartford Fire Department. He served from 1898 until his retirement in 1914, at Engine Company No. 7. Participation in the fire service by African Americans declined rapidly in the United States after 1900 due to the segregation practices. It had nothing to do with skill, dedication, or ability. Not until 1948 did this practice change in the Hartford Fire Department. (Courtesy of the Connecticut Fire Museum Archives.)

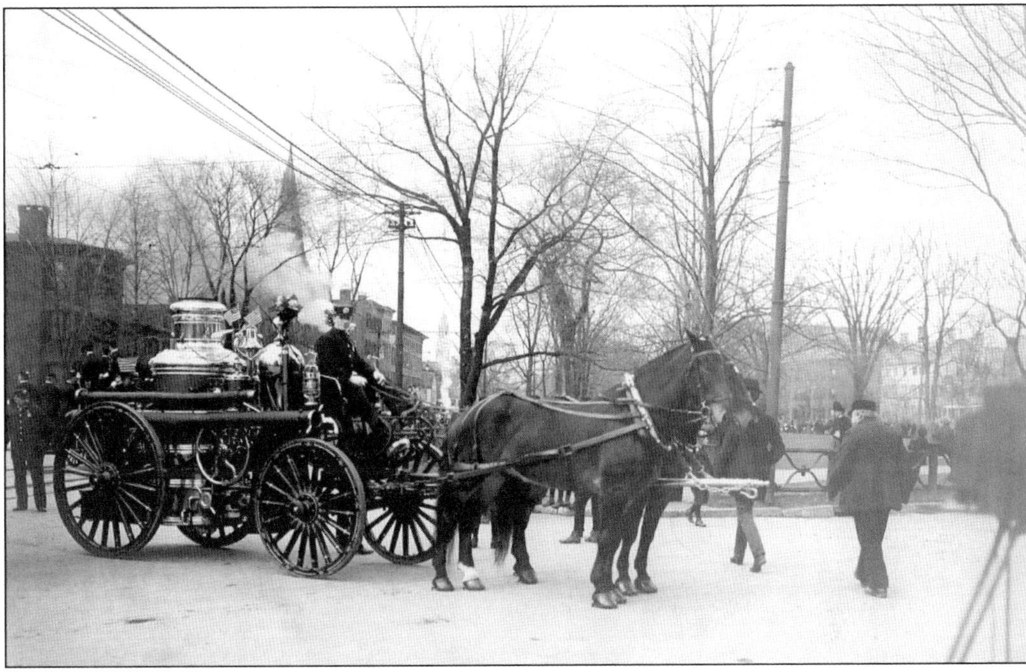

Engine No. 1's Amoskeag steam pumper, put in service in 1900, is shown in this 1905 view at the South Green. This was a "third size" pumper by the standards of the day. Today it would be called a 650-gallon-per-minute pumper. (Courtesy of the Connecticut Fire Museum Archives.)

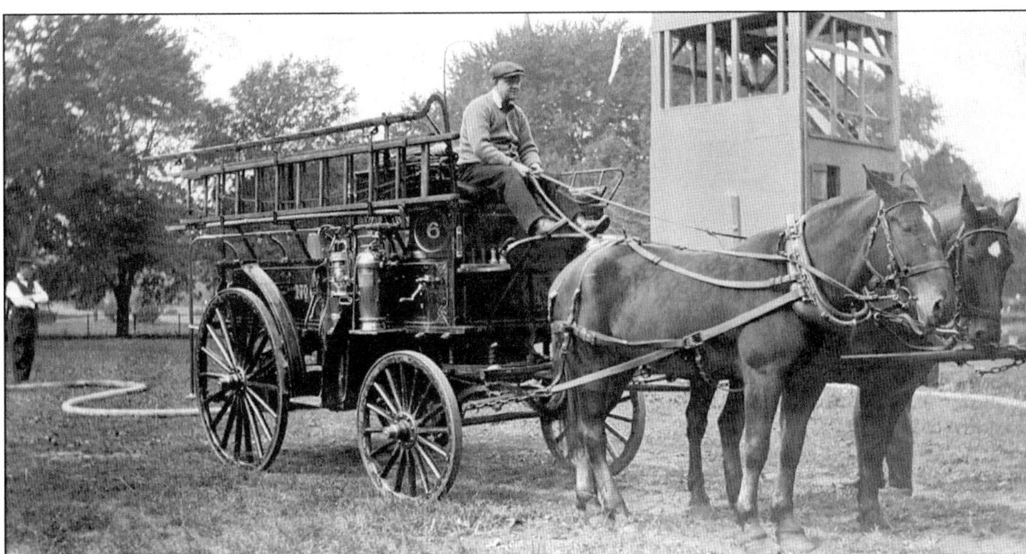

Engine Company No. 6's Holloway chemical and hose wagon is at the fire department drill tower on Huyshope Avenue in this 1910 view. A hose wagon carried ladders and firefighting tools in addition to the chemical equipment. Chemical engines originated in France to fight small fires. Upon arrival at the fire, acid is added to a soda mixture, creating a pressurized foam that consists of water and carbon dioxide (CO_2) with the mixture fed to the fire by a one-inch hose. Hose wagons and soda acid chemicals are no longer used today in fire service. (Courtesy of Mike Gorman.)

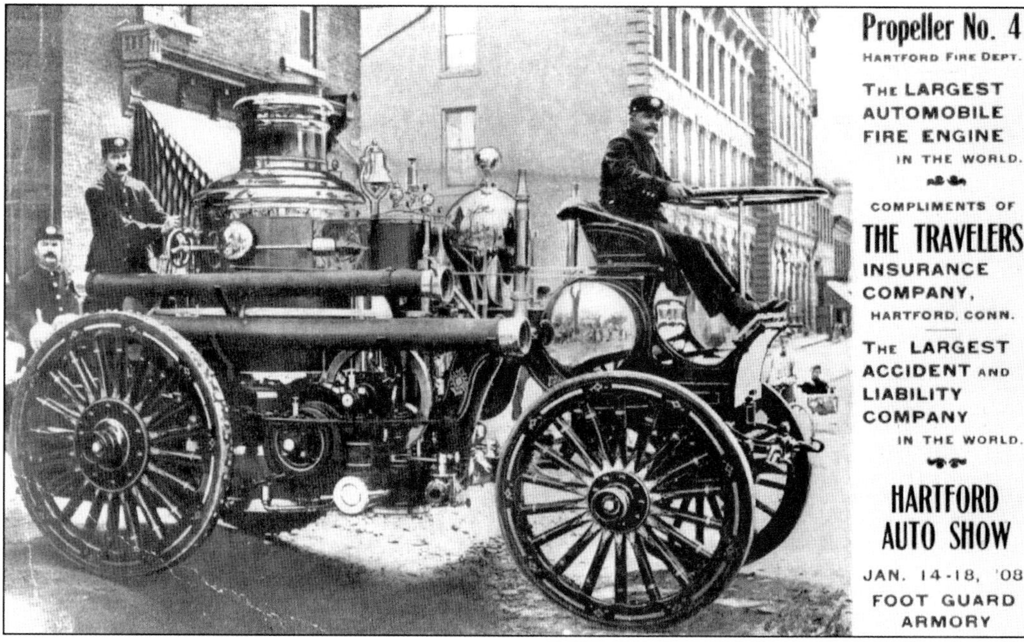

"Pride of Hartford," a self-propelled Amoskeag steam fire engine built in 1901, is shown at Engine Company No. 4's quarters on Ann Street. These engines used the steam fire pump to move the apparatus; speed and braking were controlled by the engineer on the rear platform, all the driver did was steer the engine. This engine was retired in 1925 and is now owned by the Wilson Fire Department in Windsor, Connecticut. (Courtesy of the Connecticut Fire Museum Archives.)

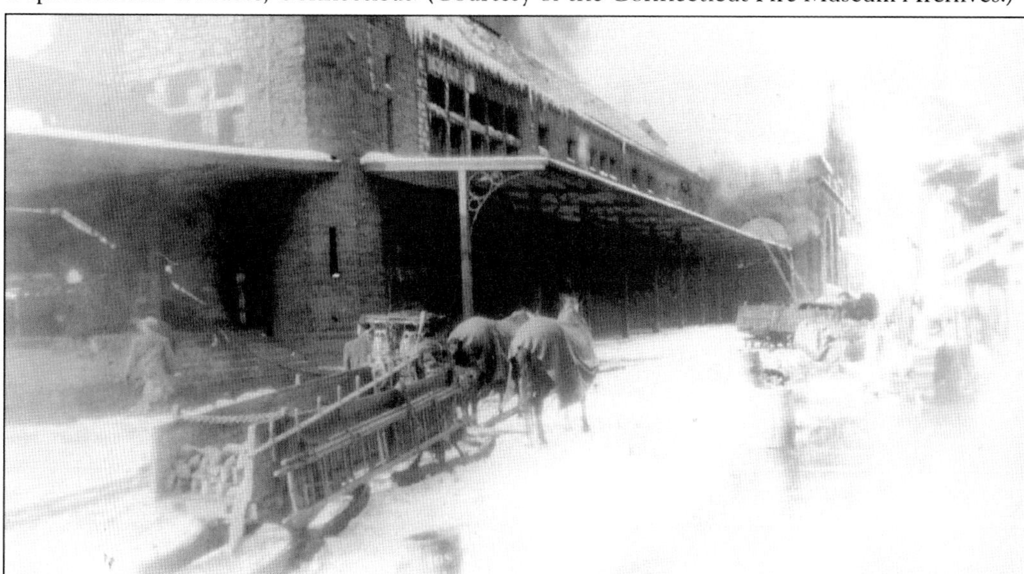

Hartford's Union Station burned on February 21, 1914, in a spectacular fire that went to three alarms with 13 Hartford companies fighting the fire. Horse-drawn sleighs were used to carry hose and equipment during times of heavy snow. Engine companies No. 1, 2, 3, 5, 10, 12, 14, and 15 all had sleighs for emergency winter use. Prior to World War I, the only streets plowed in Hartford were those that had streetcar lines, and those were plowed by the street railway company. (Courtesy of the Connecticut Fire Museum Archives.)

Three

HARTFORD MOTORIZED

In July 1912, the horse-drawn hose wagon of Engine Company No. 1 was replaced by a Pope-Hartford chemical and hose wagon at a cost of $5,259. In August, three more new Pope-Hartford units, like Engine Company No. 1's, were delivered and assigned to Engine Companies No. 2, No. 3, and No. 5. In 1914, three more combination chemical and hose wagons were received from Pope-Hartford to replace other horse-drawn hose wagons in the city. In 1914, Pope-Hartford stopped production of fire apparatus along with all its manufacturing of automobiles and trucks. In 1914, a "mutual aid region" was defined with Hartford as the central point. This arrangement meant that, in times of need, communities aided each other in fighting fires. The area was bounded by Thompsonville on the north, Rockville to the east, Meriden and Middletown to the south, and Bristol to the west. Also in 1914, an American LaFrance 1,400-gallon-per-minute pumper was ordered to motorize Engine Company No. 12. In addition, Engine Company No. 14 and Ladders No. 3 and No. 5 were equipped with motorized apparatus. During 1914, the department answered 576 alarms, including two three-alarm fires within one week: the Union Station and the auditorium building both in February. The work schedule changed in 1915 from one off in 10 days to one day off in 8 days. An agreement was signed with the East Side Fire District of West Hartford to have their Prospect Avenue engine company cover for Engine No. 12 and for Hartford to provide coverage for the East Side Fire District of West Hartford in case of a second fire. In January 1917, the G. Fox and Company building at 970 Main Street was destroyed in a general alarm fire with losses in excess of $750,000. On September 25, 1920, the horse-drawn apparatus of Engine No. 4 was replaced by motorized apparatus, thus completing the motorization of the department. The 1920s saw the arrival of "second" generation fire apparatus as the self-propeller steamers were retired and replaced by 1,000-gallon-per-minute American LaFrance apparatus. In 1920, the repair division moved into a new facility on John Street, previously the repair division occupied space at Engine No. 7's quarters on Main Street. The facility had an elevator that allowed fire apparatus to be brought up to the second floor. In 1925, all the hose, hydrants, nozzles, and other equipment used by the department were refitted with National Standard thread to be compatible with other departments in the mutual aid area around Hartford. By the early 1930s, the department had 15 engine companies, six ladder companies, and one water tower, which operated with eight 1,000-gallon-per-minute pumpers, six 750-gallon-per-minute pumpers, and one 500-gallon-per-minute pumper. A new mutual aid agreement was drawn up in 1932 with Hartford, East Hartford, West Hartford East Side, West Hartford Center, New Britain, Newington, Wethersfield, Bloomfield, South Manchester, and Rockville all party to the agreement. Hartford notified all towns party to the agreement that a charge of $50 per hour for each company called, from the time it left the station until it was back in service, would be charged to the town requesting mutual aid assistance. During the 1930s, the city purchased a number of Mack engines and tractors; this was a change in the long-standing policy of standardization, which usually meant that American LaFrance would get the order for the equipment. In 1937, a mutual aid agreement, similar to the agreement with the East Side Fire District of West Hartford, was signed with the Blue Hills Fire District of Bloomfield after they purchased a new pumper similar to Hartford equipment.

On June 1, 1937, Chief John C. Moran officially retired from the department; he had actually been on paid leave since December 1, 1936. Michael T. Keene succeeded him. In December 1941, the department assisted the Glastonbury Fire Department when the Angus Park woolen mill in East Glastonbury burned. In the 1940s, Hartford was averaging about 1,400 runs per year. On July 24, 1943, Hartford firefighters were unionized when Local 760 International Association of Firefighters was organized. July 6, 1944, was the date of the tragic Barnum and Bailey Circus fire on Barbour Street. During the 2:00 p.m. show, the large main tent erupted in flames. By the time the fire was out, 168 people were dead and many more were injured. Hartford in the 1940s was a "rolling museum" of antique fire apparatus with many companies using apparatus from the 1920s. With the end of World War II, in August 1945, the fire apparatus manufacturers were gearing up for the flood of orders for new equipment that could not be ordered by the cities because of the wartime restrictions. American LaFrance and Mack Trucks would be the builders of choice.

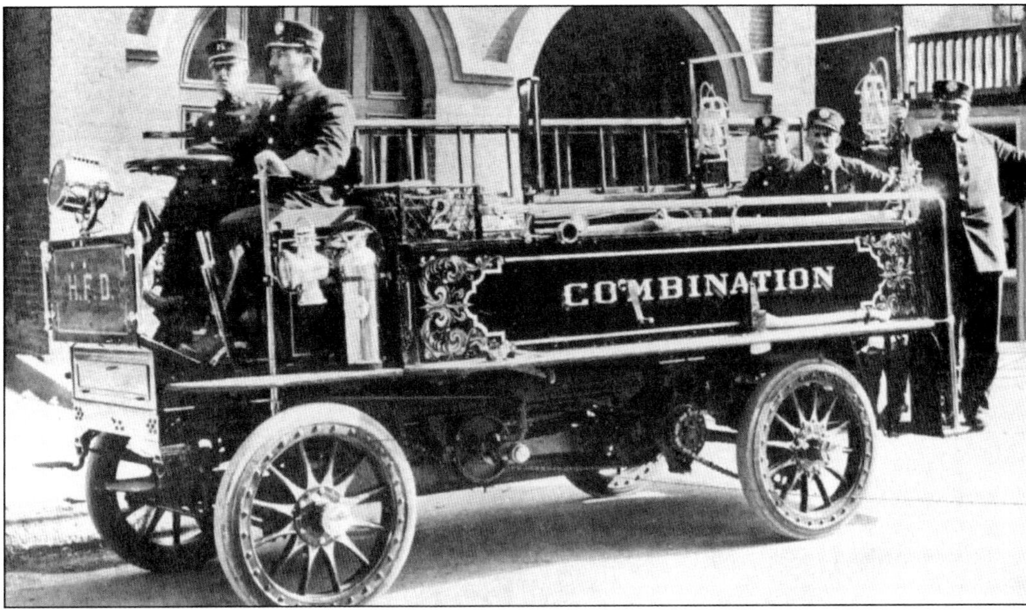

This is a 1907 Knox chemical and hose wagon. This unit had a 40-horsepower opposed cylinder engine under the driver seat. The Combination Ladder Company of Providence, Rhode Island, built the fire engine body for Knox. The maximum speed was 20 miles per hour, and it could carry 1,000 feet of two-and-a-half-inch hose and a crew of five. The driver's seat above the engine is a carryover from the horse-drawn days This unit was in service until the late 1930s. (Courtesy of Mike Gorman.)

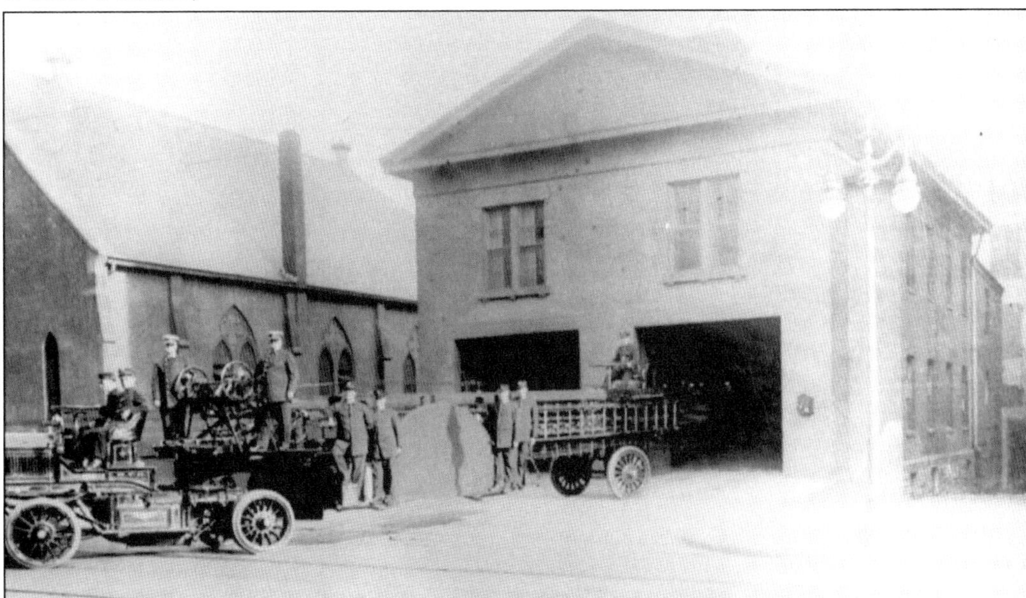

Ladder Company No. 1's 1914 American LaFrance 85-foot tillered aerial truck is shown in this 1915 scene at 275 Pearl Street. The present fire headquarters building occupies the site of the church and firehouse shown in this scene. American LaFrance delivered its first motorized fire apparatus in 1910 to Lenox, Massachusetts, which still owns and operates the truck in 2006! (Courtesy of Mike Gorman.)

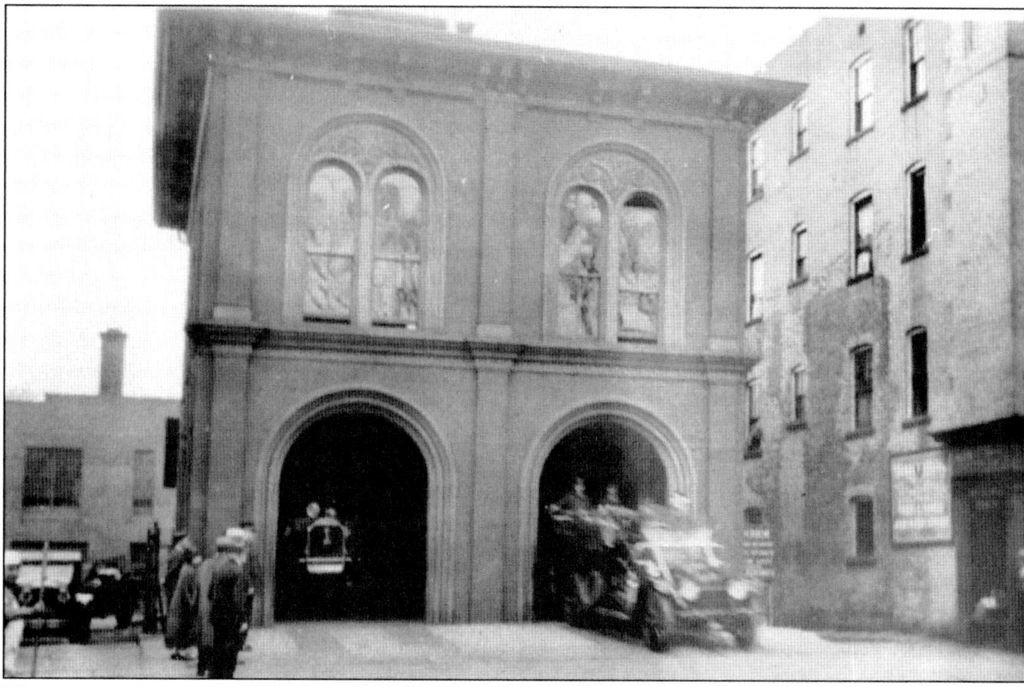

This image shows Engine Company No. 1's quarters at 197 Main Street around 1918. (Courtesy of Mike Gorman.)

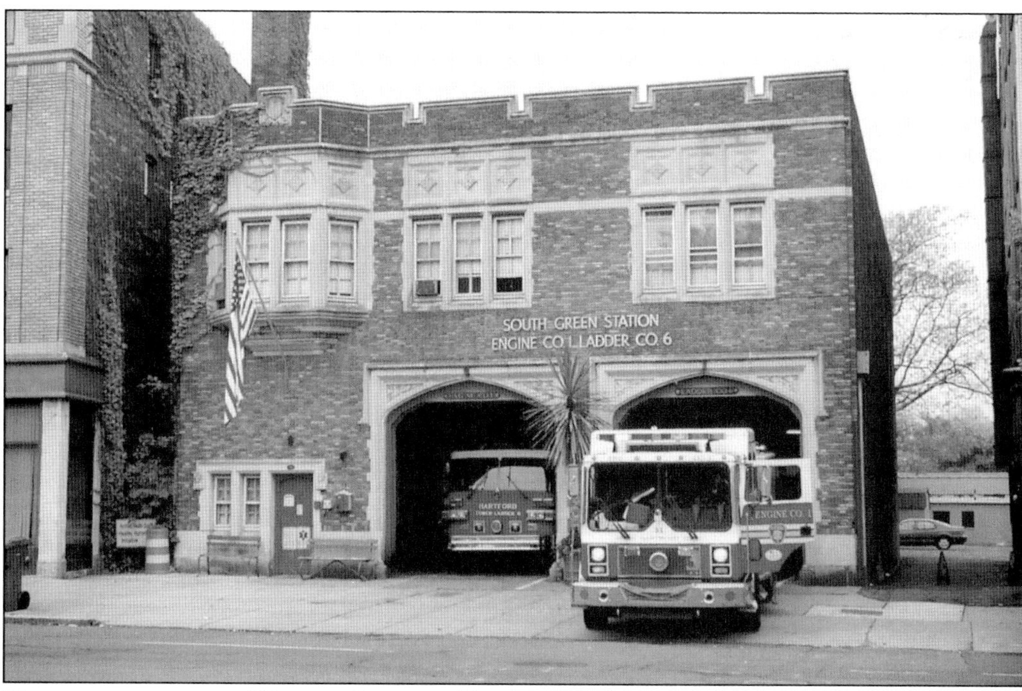

The present quarters of Engine Company No. 1 and Ladder Company No. 6 was built in 1926, on the same site as the building on the top of this page. (Courtesy of the Connecticut Fire Museum Archives.)

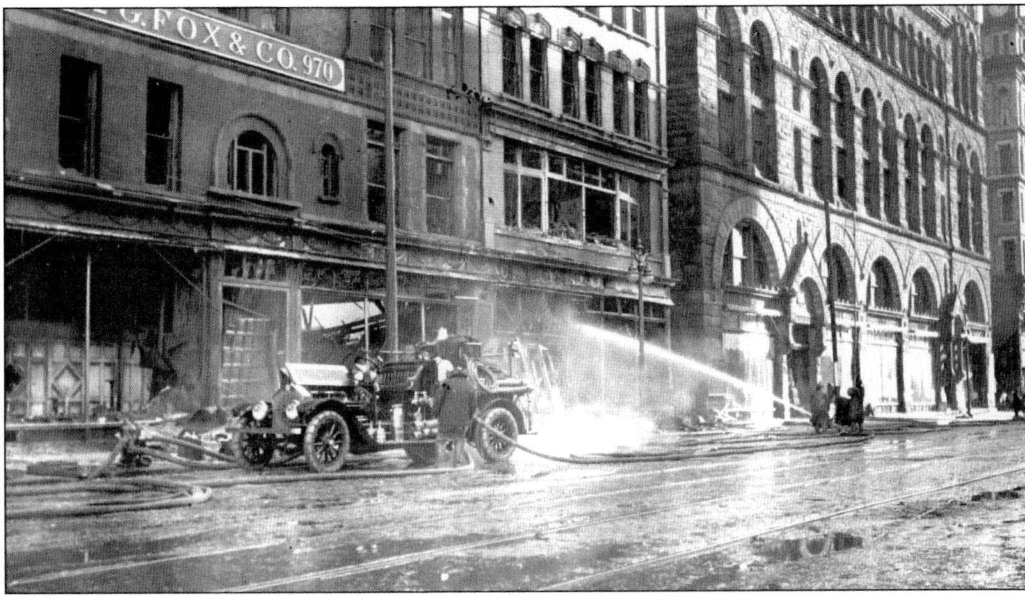

The G. Fox and Company store at 970 Main Street was destroyed by fire on January 29, 1917. This was the most serious fire in Hartford under the paid department up to that time, with a loss of over $750,000. It took the entire department to bring the fire under control. The East Side Fire District of West Hartford as well as the New Britain and South Manchester Fire Departments provided mutual aid to Hartford. (Courtesy of Mike Gorman.)

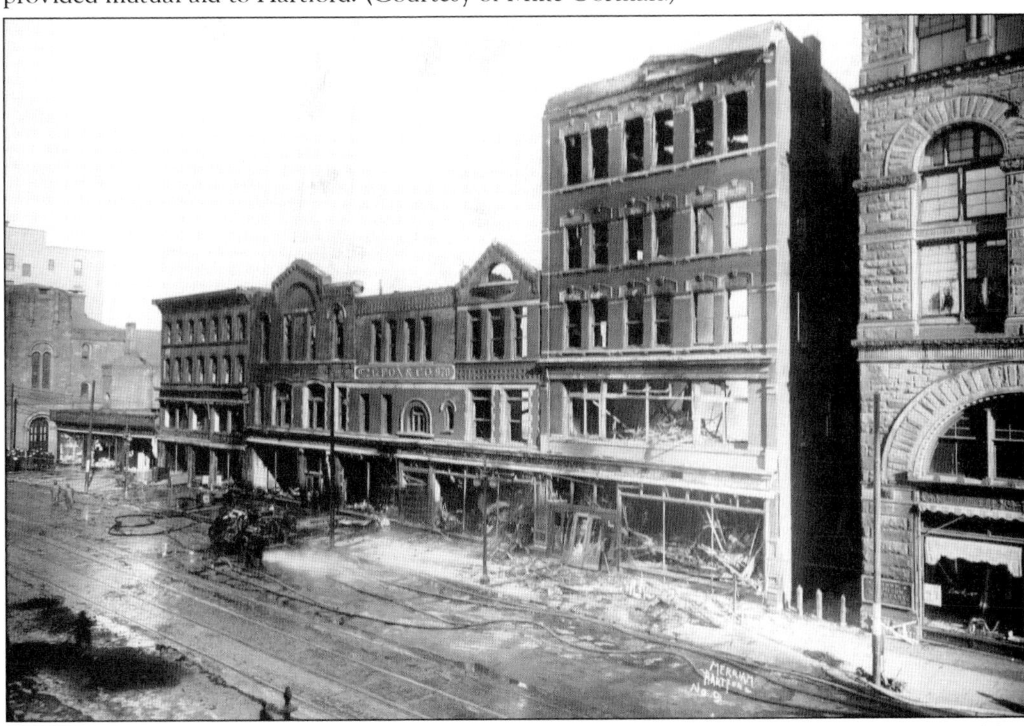

Here is a scene from the day after the fire, which shows the damaged building that was a total loss. Within a few years, G. Fox and Company reopened a new store on the same site. This building still stands today and is home to Capital Community College. (Courtesy of Mike Gorman.)

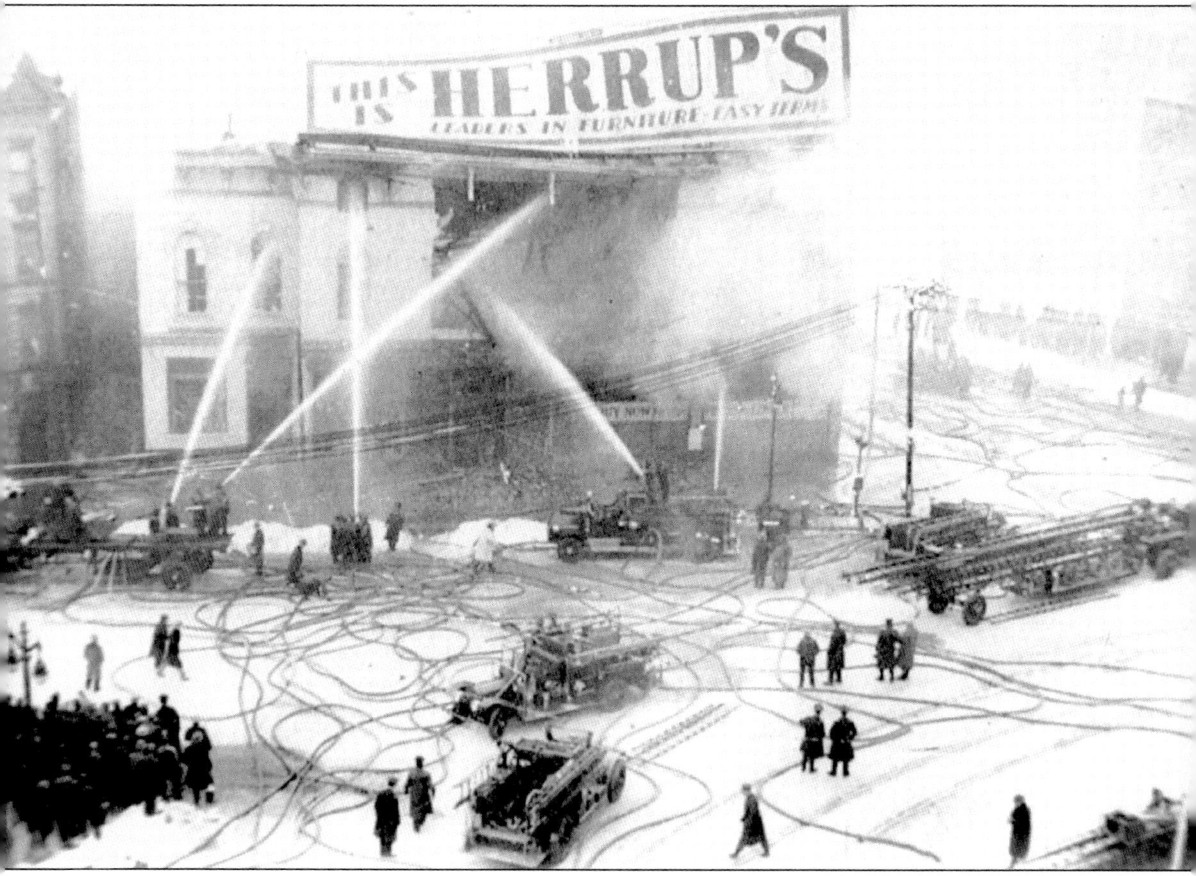

Herrup's Furniture Store at 1052 Main Street was destroyed by fire on February 6, 1932. In 1931, the National Board of Underwriters surveyed the department and city and reported that the department was considered well organized, well equipped, and led by experienced and progressive officers. Fire insurance rates stayed low as a result of this survey. Staffing, apparatus, fire hydrants, the fire alarm system, and training were all evaluated by the underwriters when doing their study. (Courtesy of Mike Gorman.)

Engine Company No. 10 was organized in 1901, and its first station was located at 94 Bond Street. In this 1914 scene is the Pope-Hartford chemical and hose wagon assigned to the company. In 1961, the company moved into a new station at 529 Franklin Avenue. (Courtesy of Mike Gorman.)

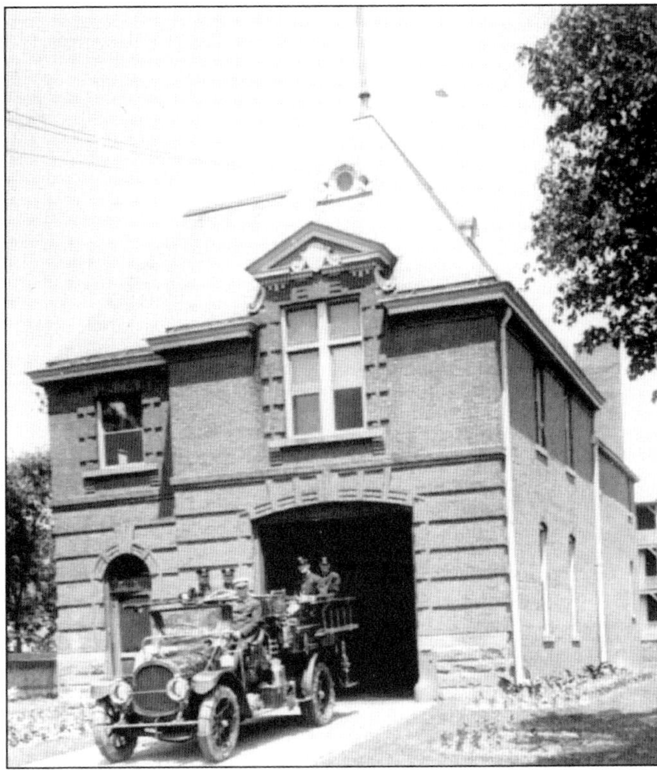

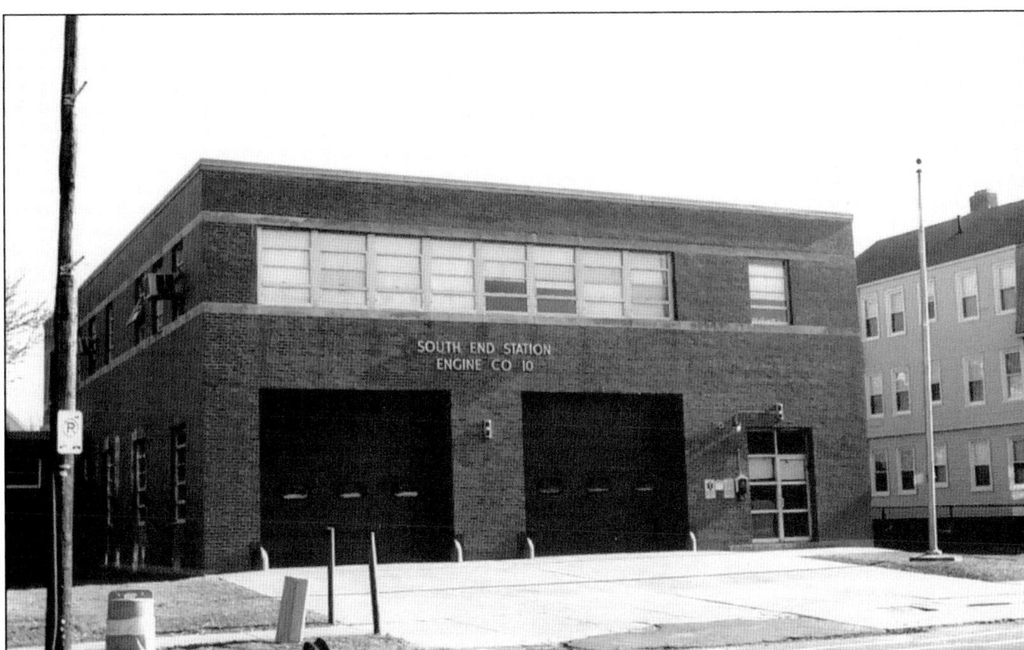

This is a 2006 view of Engine Company No. 10's quarters on Franklin Avenue. (Courtesy of the Connecticut Fire Museum Archives.)

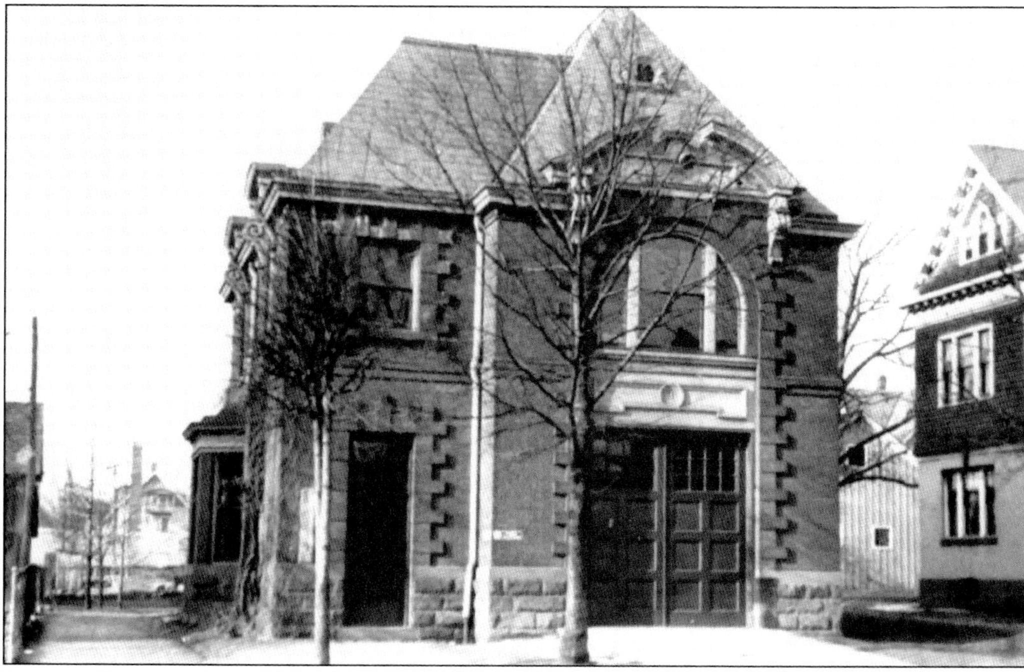

Engine Company No. 11's station at 3 Sisson Avenue is shown in this 1920 view. Engine Company No. 11 was organized in 1901 as Chemical No. 11. In 1961, this station was closed when Engine No. 11, Engine No. 12, and Ladder No. 5 were all relocated to a new station at 150 Sisson Avenue. (Courtesy of Mike Gorman.)

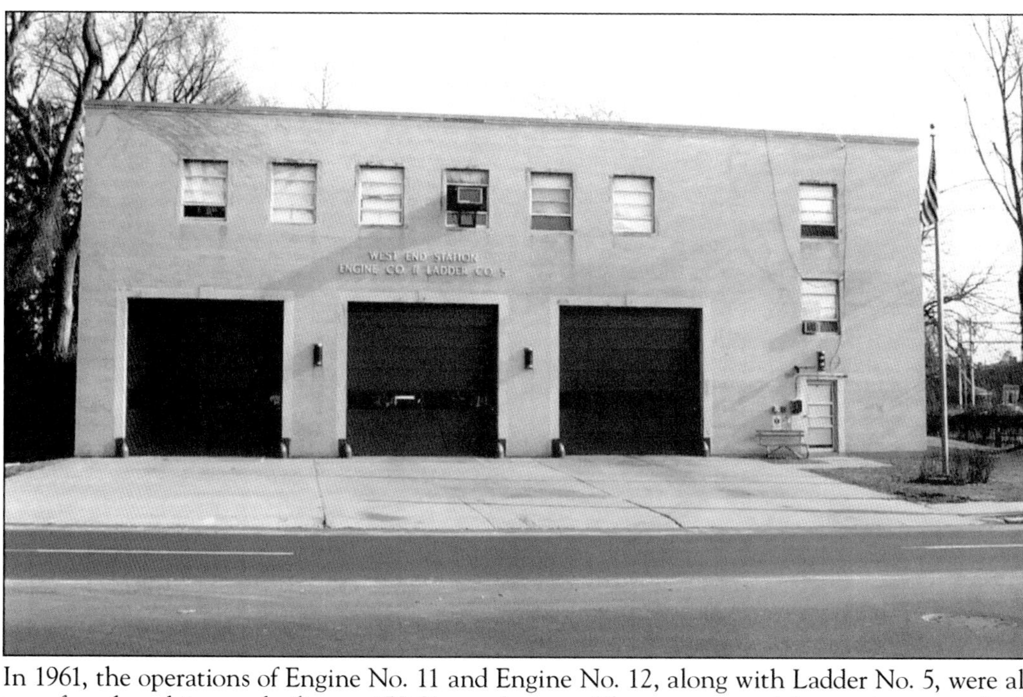

In 1961, the operations of Engine No. 11 and Engine No. 12, along with Ladder No. 5, were all transferred to this new facility at 150 Sisson Avenue. The two vacated fire stations were both torn down and the property sold. (Courtesy of the Connecticut Fire Museum Archives.)

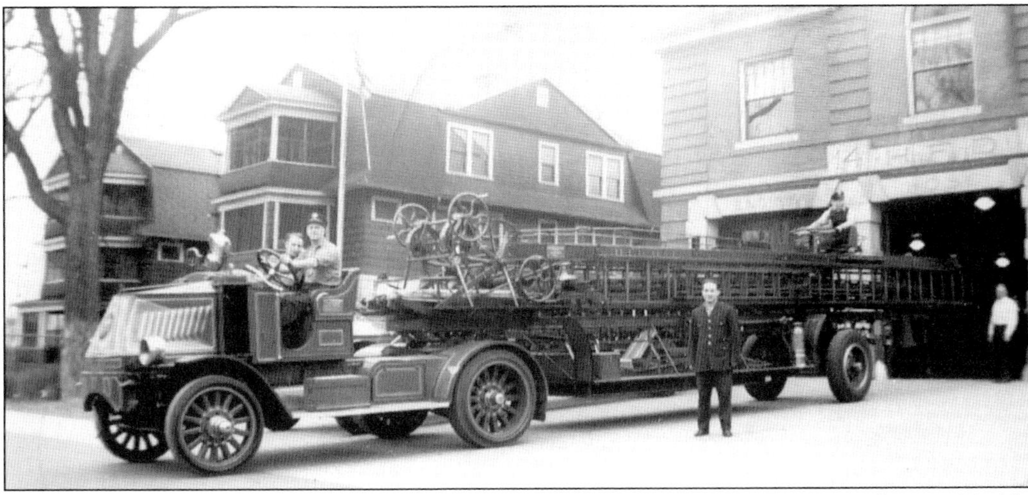

Charles Argard is at the wheel of Ladder Company No. 4's 1923 Mack. One can see the aerial ladder with driver at Engine Company No. 14 on Blue Hills Avenue. (Courtesy of Mike Gorman.)

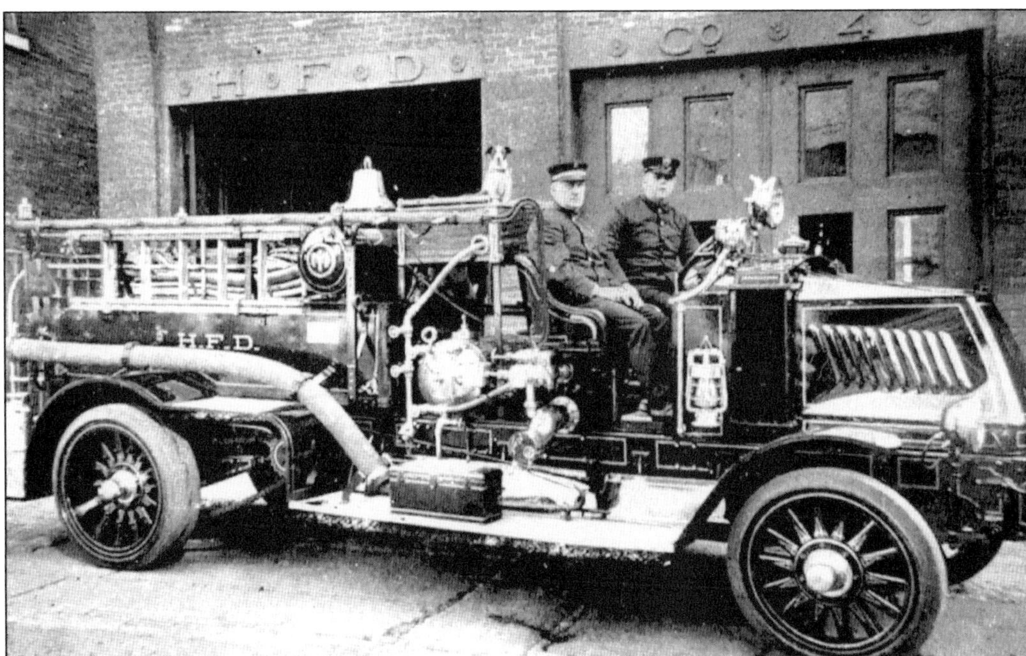

This is Engine Company No. 4's 1923 Mack pumper with its crew and the firehouse dog Skippy. This truck later went to Engine Company No. 16 on Blue Hills Avenue when it opened in 1928. Apparatus of this era used hand-cranked sirens; however, the main sound for alerting the public was the apparatus themselves, as they had no mufflers and made a very loud roaring noise. (Courtesy of Mike Gorman.)

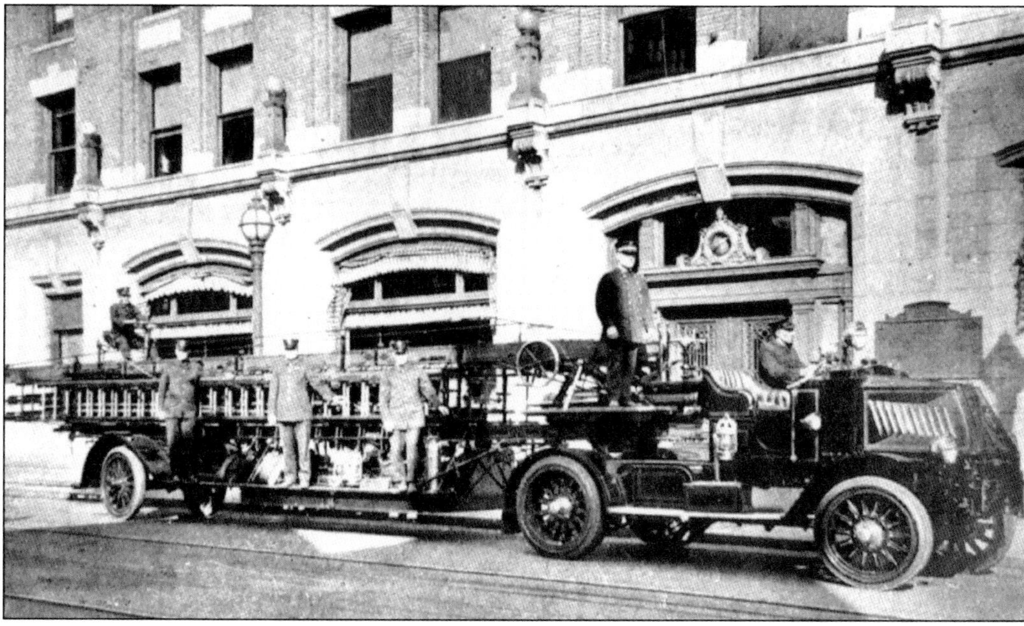

Here is another view of the 1923 Mack aerial ladder truck, this time taken in the late 1920s when it was assigned to Ladder Company No. 1 on Pearl Street. The building behind the truck is the home office of the Hartford Electric Light Company located across the street from fire headquarters at 275 Pearl Street. (Courtesy of Mike Gorman.)

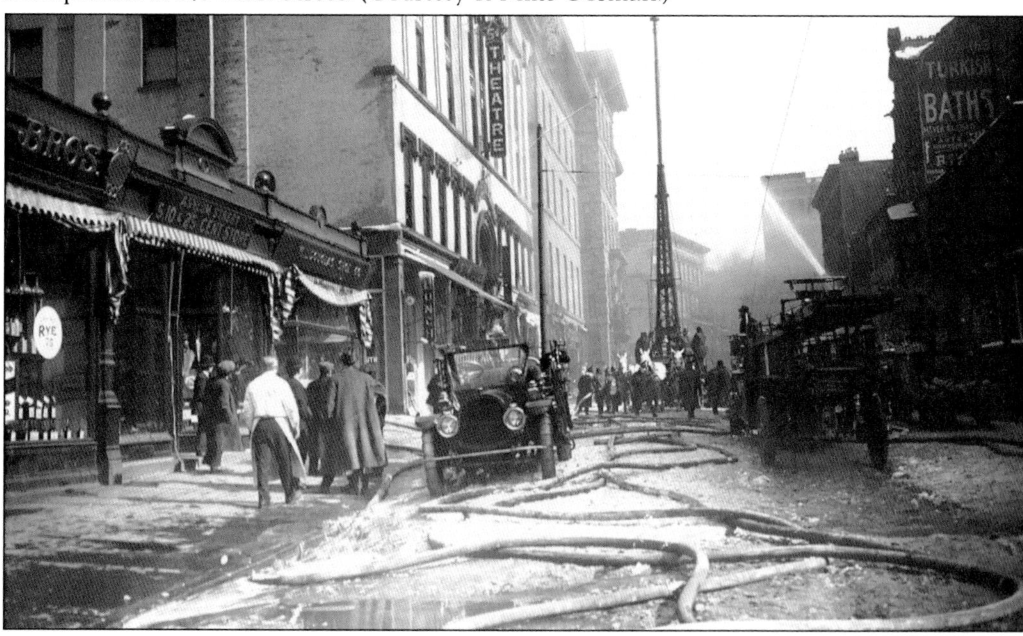

On February 26, 1914, the auditorium building at 183 Asylum Street burned in a three-alarm fire that resulted in over $50,000 in damages. In this view, one of the Pope-Hartford chemical and hose wagons and the water tower, which is still horse drawn, can be seen. The three horses are still hitched to the water tower, so this picture was taken just after the equipment arrived at the scene. They will soon be unhitched and walked to cool them off, or be covered with blankets if the weather was cold. Fire horses always received loving care. (Courtesy of Mike Gorman.)

This is an aerial view of the auditorium building fire in its early stages. (Courtesy of Mike Gorman.)

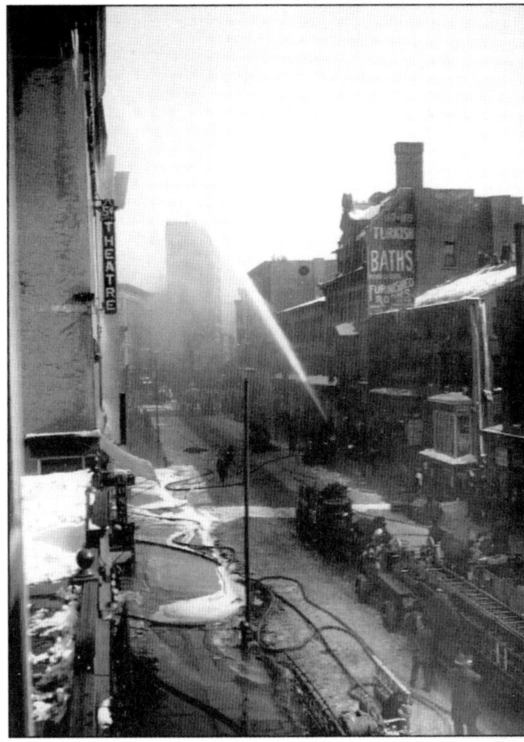

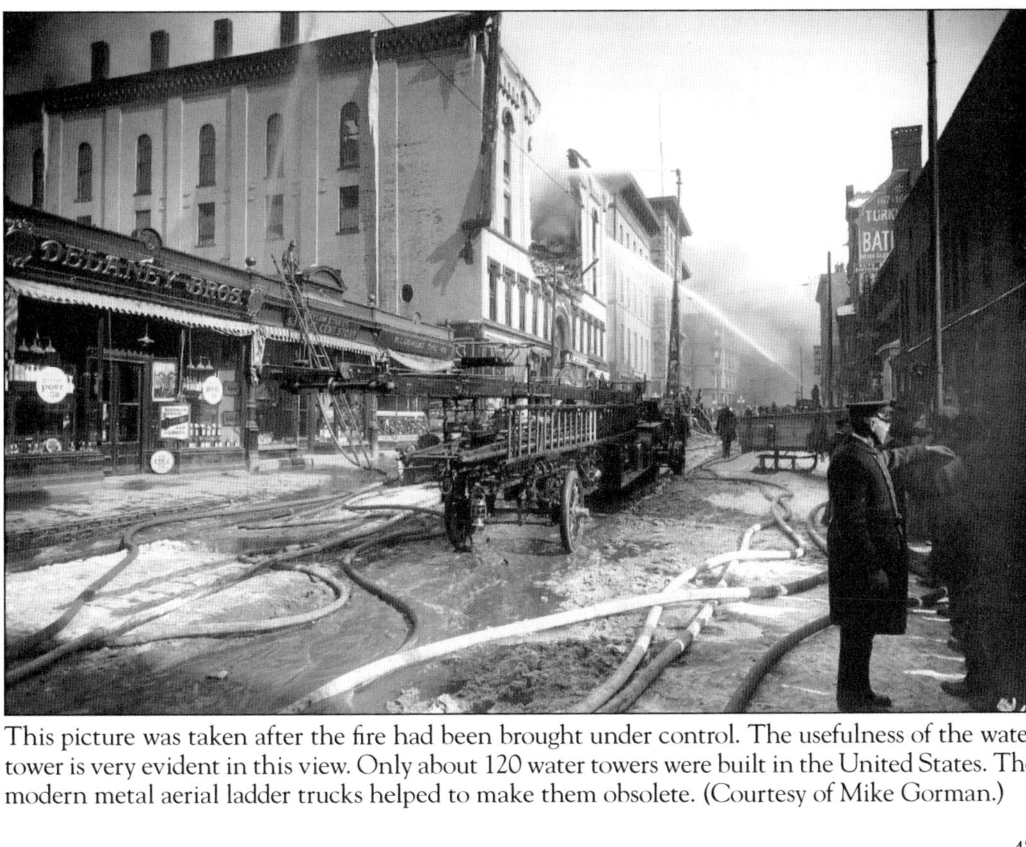

This picture was taken after the fire had been brought under control. The usefulness of the water tower is very evident in this view. Only about 120 water towers were built in the United States. The modern metal aerial ladder trucks helped to make them obsolete. (Courtesy of Mike Gorman.)

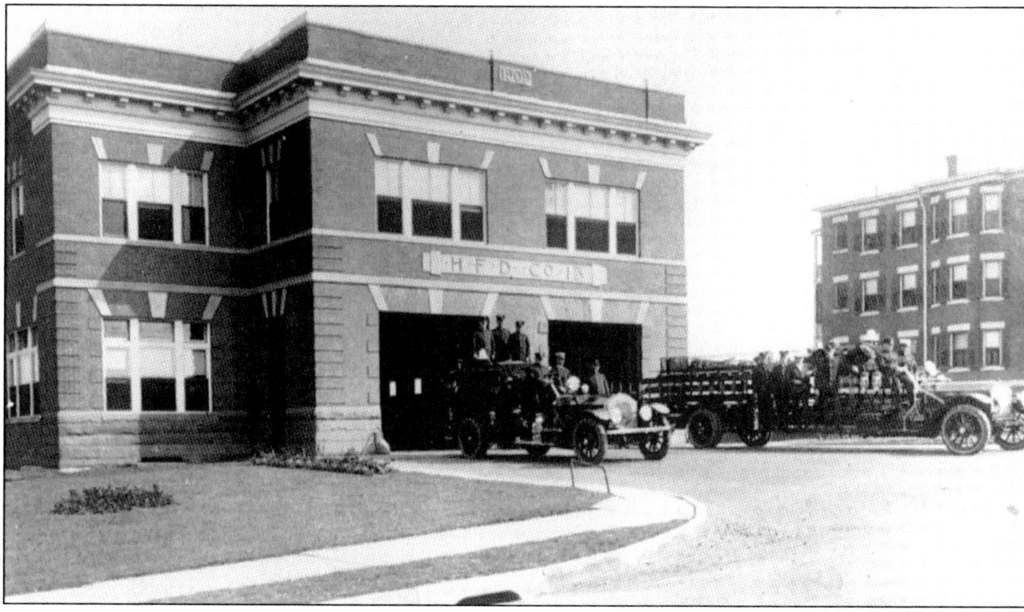

Station No. 15, located at New Britain and Fairfield Avenues, is shown in this *c.* 1916 photograph with Engine No. 15, a 1915 American LaFrance type 12, 750-gallon-per-minute pumper, and Ladder No. 2, a 1915 American LaFrance cities service ladder truck. Cities service trucks only carried ground ladders and did not have an aerial ladder. This station opened in 1909 and is still in service in 2006. Engine Company No. 15 was closed briefly in 1991 due to a manpower shortage. (Courtesy of Mike Gorman.)

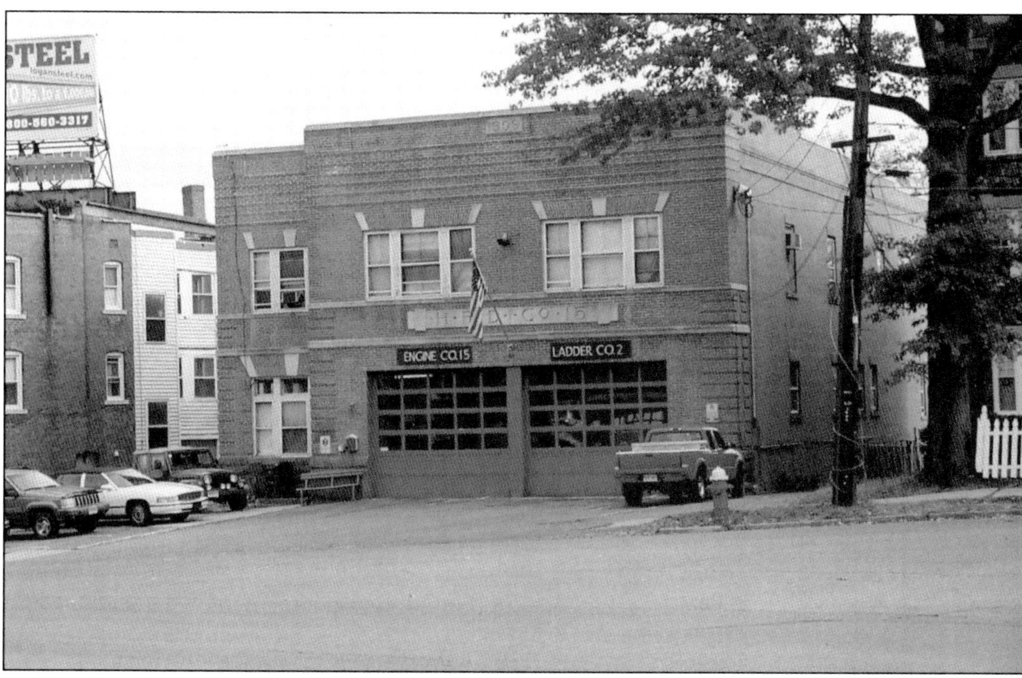

This is a 2006 view of Engine Company No. 15's station. (Courtesy of the Connecticut Fire Museum Archives.)

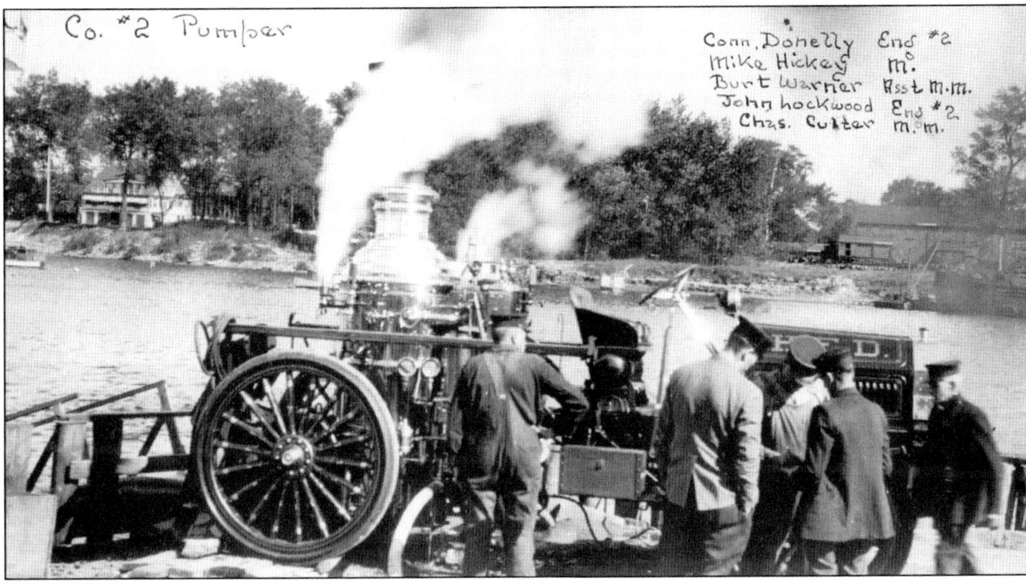

Every spring, the fire department tested all the pumpers at a site on the Connecticut River. In this view is Engine Company No. 2's American LaFrance steam pumper drafting from the river. Observing the operation is the staff from the mechanical department and members of Engine Company No. 2. (Courtesy of Mike Gorman.)

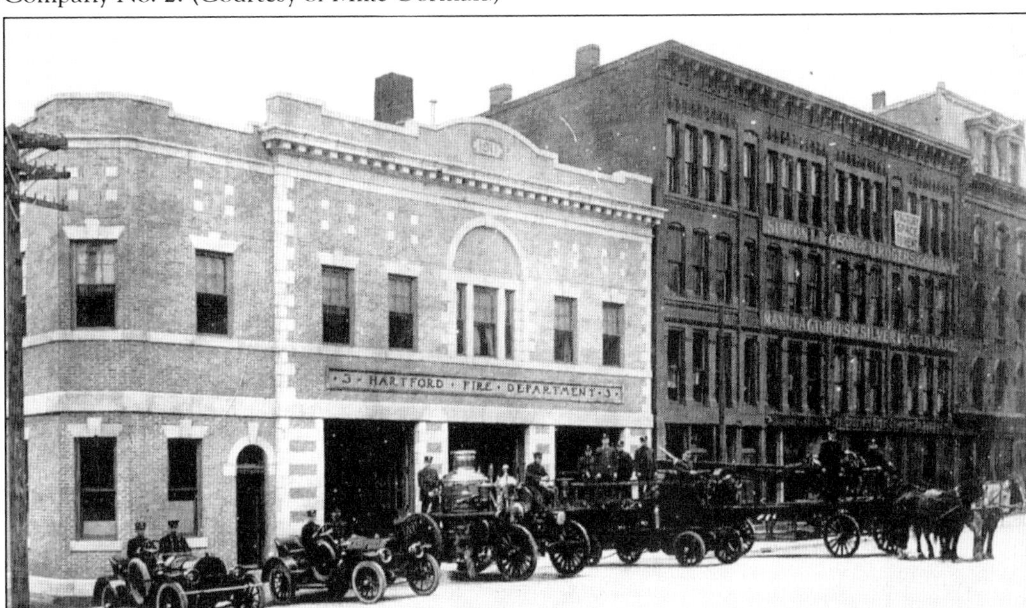

Engine Company No. 3's station at 90 Market Street is shown in this c. 1915 photograph with the department's two Pope-Hartford automobiles used by the chief and his assistants. "Jumbo," the self-propeller steamer, a 1907 Knox hose wagon, and the 1911 American Automatic water tower all operated out of this station. The station was closed in 1958 and demolished for construction of Constitution Plaza. Engine No. 3 was briefly housed with Engine No. 4 on Pearl Street. In 1961, Engine No. 3 relocated to a new station at 182 Clark Street that was shared with Engine No. 7. (Courtesy of Mike Gorman.)

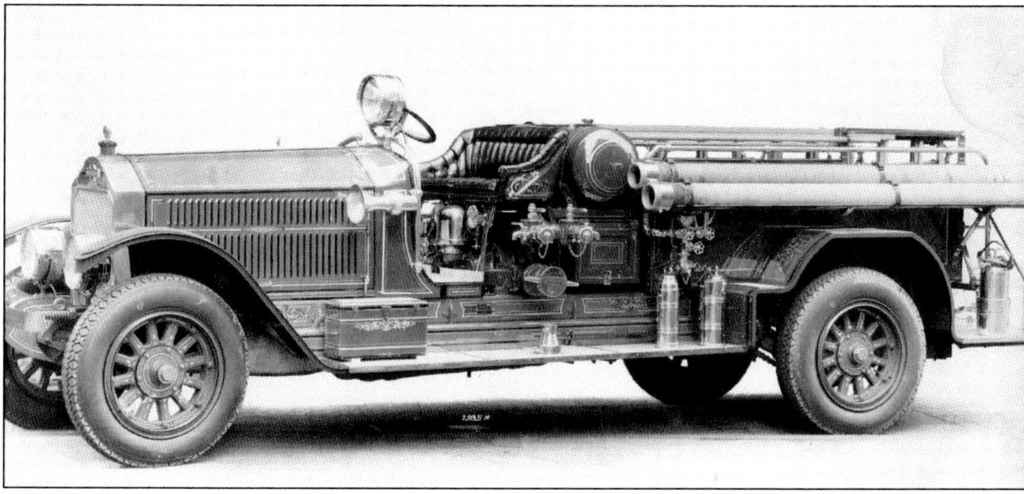

Hartford purchased two 1,000-gallon-per-minute pumpers from American LaFrance in 1928, like the one in this view. Originally delivered with a chemical tank, these were removed by the department in the 1940s, replaced with a 90-gallon water tank. Engine Company No. 7's unit, assigned shop designation "car 36," was the first unit to arrive at the circus fire on July 6, 1944. This truck had been preserved and is now on display at the Connecticut Fire Museum in East Windsor. (Courtesy of Mike Gorman.)

Fireman's Pay

In 1928 the typical work week for a full time firemar in Hartford was 84 hours.

Yearly Salaries

Position	Annual Salary
Chief	$3900
Assistant Chief	3120
District Chief	3060
Fire Marshall	3000
Superintendant of Fire Alarms	3000
Master Mechanic	3000
Assistant Master Mechanic	2640
Senior Driver	2580
Junior Driver`	2520
Lineman	2400
Class B Fireman	1800
Class C Fireman	1800
Class D Fireman	1680

Here one can see fireman pay rates in 1928. (Courtesy of the Connecticut Fire Museum Archives.)

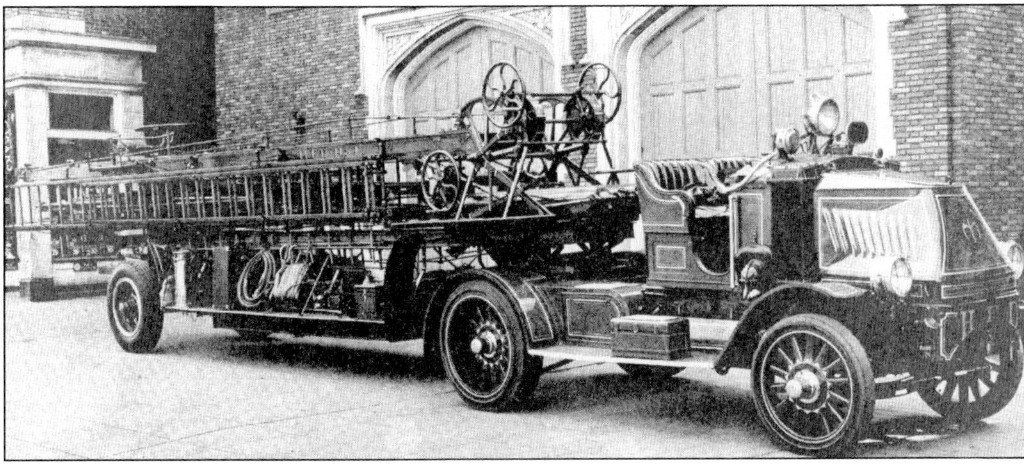

Ladder Company No. 1's 1923 Mack 85-foot aerial ladder is shown in front of Engine No. 1's quarters in 1923. The Hartford Fire Department machine shop photographed all new apparatus when delivered to the city. (Courtesy of Mike Gorman, James Callahan Collection.)

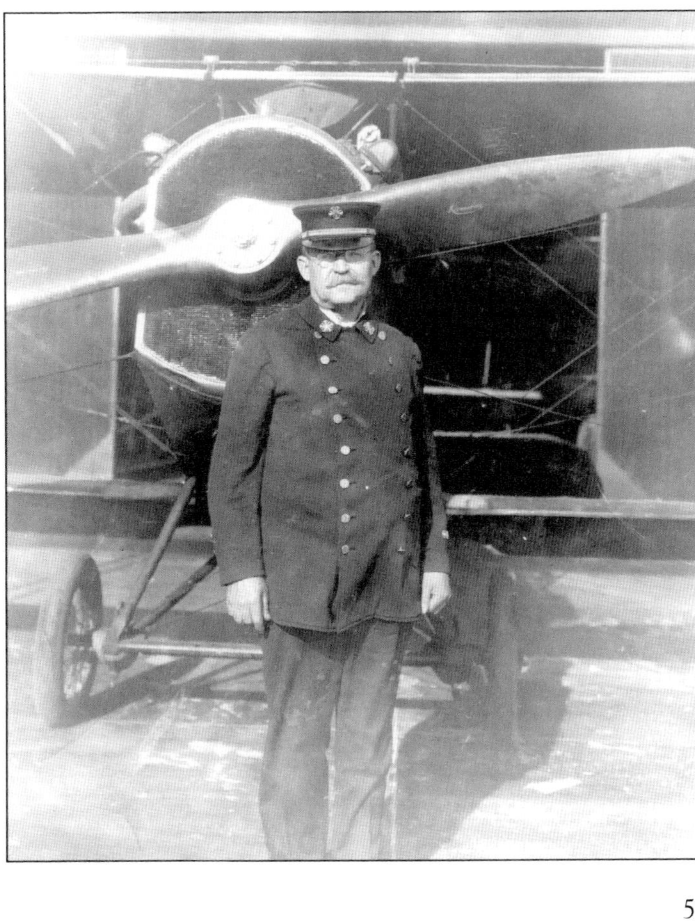

John C. Moran served on the fire department for over 50 years, serving as chief from 1913 until he officially retired on June 1, 1937. Prior to becoming chief, he had served as engineer and a captain of Engine Company No. 4. (Courtesy of Mike Gorman, John D. Jansen Jr. Collection.)

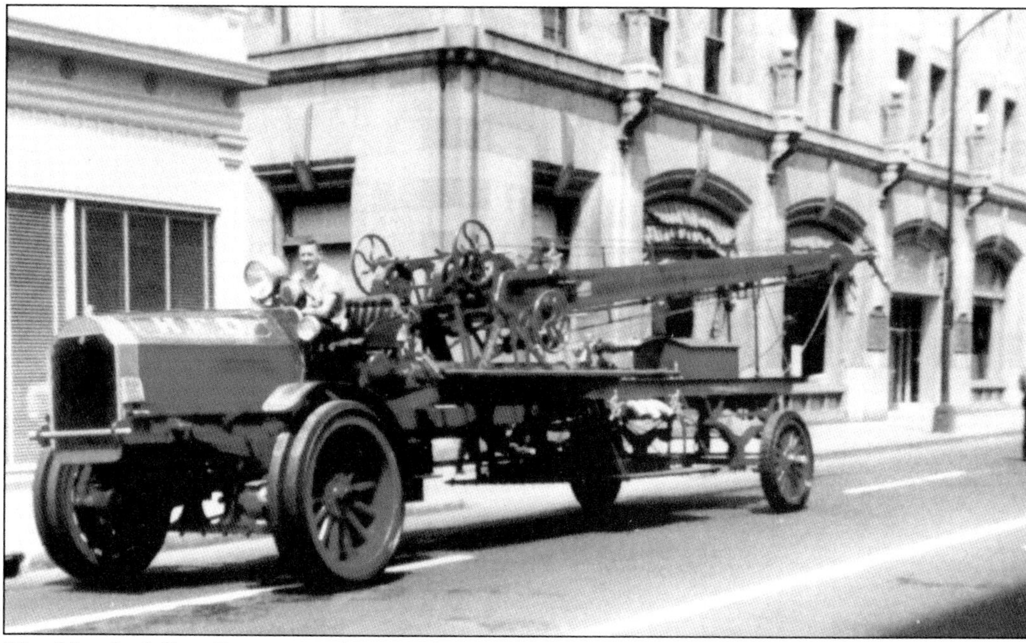

The Hartford Fire Department's 1911 American Automatic water tower was in service until the 1950s. It can be seen today at the Connecticut Firemen's Historical Society Museum in Manchester. (Courtesy of Mike Gorman.)

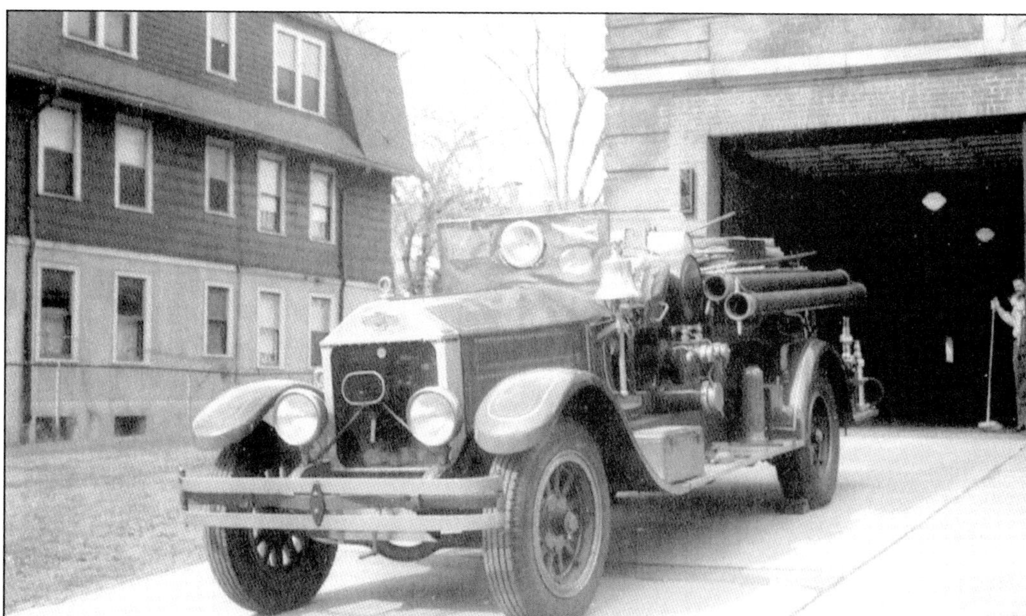

This 1928 American LaFrance 1,000-gallon-per-minute pumper was the first of the "second generation" apparatus on the Hartford Fire Department and the last with right-hand drive. It is shown in this 1968 view at Engine No. 14's quarters, where it served as a spare until the late 1970s. These trucks were built from scratch to the buyer specifications. They were some of the best engines built in the world at the time. Trucks like this typically cost over $10,000 in 1928. (Courtesy of the Thomas E. Holcombe Collection.)

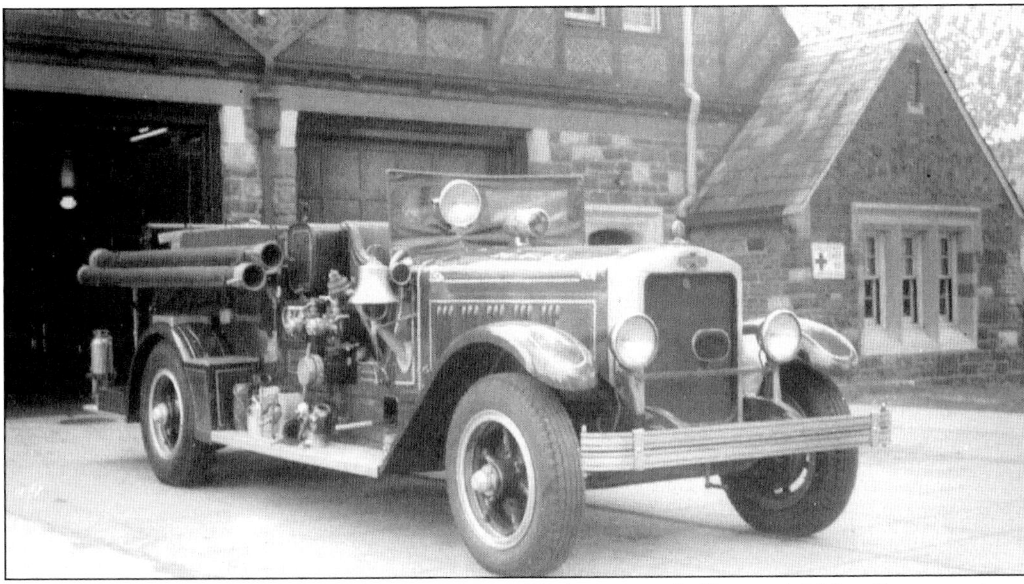

This photograph shows Engine No. 16's spare 1930 American LaFrance 1,000-gallon-per-minute pumper. Hartford purchased four of these Series 200 pumpers. This was the first year American LaFrance featured left-hand drive, four-wheel brakes, and direct shaft drive. Prior to 1930, American LaFrance preferred chain drive instead of shaft drive. Gas mileage was not great, but that really did not matter when the truck typically only operated a few miles per day. Three of these units served as spares well in the 1970s. (Courtesy of the Thomas E. Holcombe Collection.)

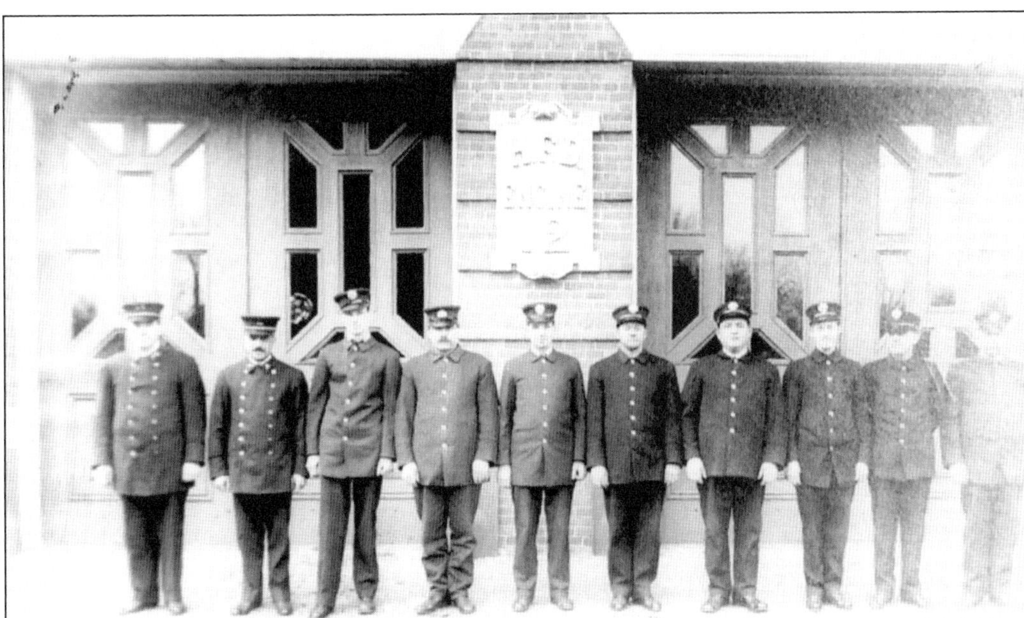

Members of Engine Company No. 12 in the 1920s are shown in this view. Prior to 1915, members of the department worked a schedule that called for a day off very 10 days. After 1915, the schedule was modified to allow for one day off every eight days. Again in 1940, the schedule was modified to allow for one off in seven. (Courtesy of Mike Gorman.)

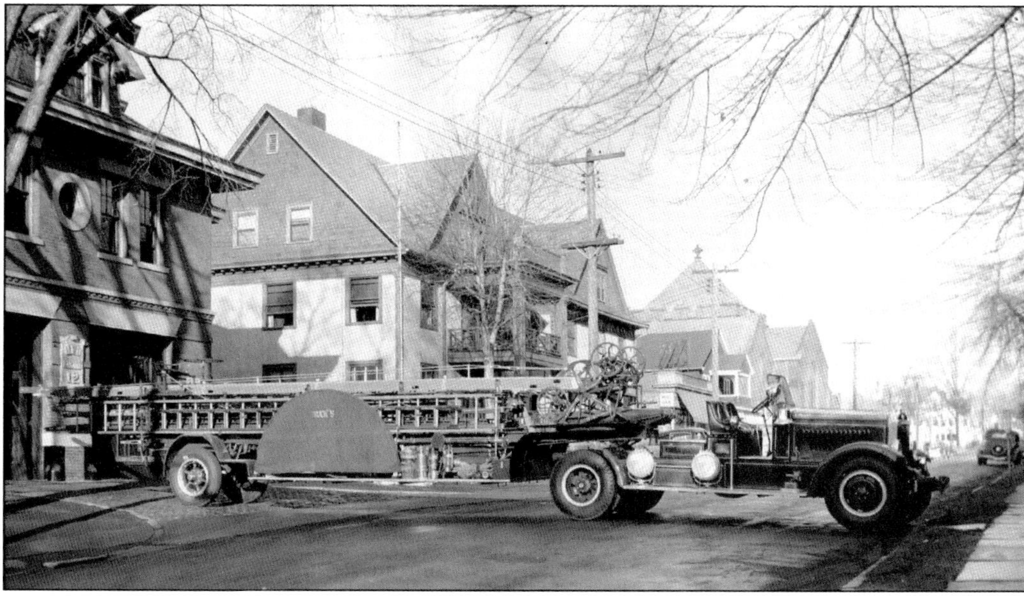

Ladder Company No. 5's 1934 Mack tractor and its 1910 American LaFrance 85-foot aerial ladder is shown in this 1940s view. The aerial ladder was originally horse drawn. The horses were originally replaced by a 1914 American LaFrance tractor. This truck was a spare unit until the 1970s and housed with Engine Company No. 10 on Franklin Avenue. Jack Jansen, a member of the Hartford Fire Department, purchased this truck in the 1970s. (Courtesy of Mike Gorman.)

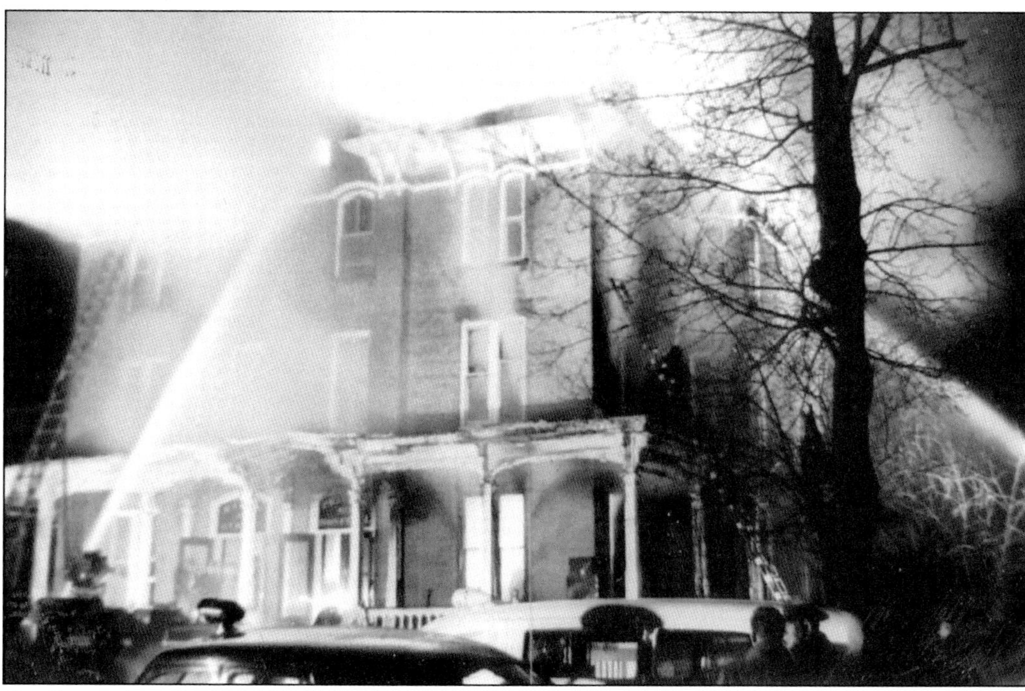

On December 24, 1944, a three-alarm fire destroyed the Niles Street Convalescent Hospital, with 21 people losing their lives. Defective wiring was cited as the cause of the fire. (Courtesy of Mike Gorman.)

1937
Stations and Apparatus

Engine Co. # 1*	1912 Pope Hartford Chemical & Hose Car
	1929 American LaFrance 1000 gpm pumper
Ladder Co. # 6	1923 Mack 85' tiller aerial
197 Main Street	
Engine Co. #2*	1914 American LaFrance 750 gpm pumper
	1930 American LaFrance 1000 gpm pumper
Ladder Co. #3	1914 American LaFrance 65' tiller aerial
Main and Belden Streets	
Engine Co. #3*	1920 Mack Chemical & Hose Car
	1930 American LaFrance 1000 gpm pumper
90 Market Street	
Engine Co. #4*	1915 American LaFrance 750 gpm pumper & hose car
	1934 American LaFrance 1000 gpm pumper
Ladder Co. #1	1928 American LaFrance 85' tiller aerial
Water tower #1	1911 American Automatic with 1919 American-LaFrance tracto
Lighting Wagon #1	1931 Ford
275 Pearl Street	
Engine Co. #5	1914 Pope-Hartford Chemical & Hose Car
	1928 American LaFrance 1000 gpm pumper
129 Sigourney Street	
Engine Co. #6	1925 American-LaFrance 1000 gpm pumper
98 Huyshope Avenue	
Engine Co. #7	1928 American-LaFrance 1000 gpm pumper
2384 Main Street	
Engine Co. #8*	1911 Pope-Hartford Chemical & Hose Car
	1930 American-LaFrance 1000 gpm pumper
721 Park Street	
Engine Co. #9	1916 American-LaFrance 750 gpm pumper
655 New Britain Avenue	
Engine Co. #10	1917 American-LaFrance 750 gpm pumper
94 Bond Street	
Engine Co. #11	1916 American-LaFrance 750 gpm pumper
3 Sisson Avenue	
Engine Co. #12	1930 American-LaFrance 1000 gpm pumper
Ladder Co. #5	1914 American-LaFrance 75' tiller aerial
243 South Whitney Street	
Engine Co. #14	1935 American-LaFrance 750 gpm pumper
Ladder Co. #4	1914 American-LaFrance cities service ladder truck
25 Blue Hills Avenue	
Engine Co. #15	1915 American-LaFrance 750 gpm pumper
Ladder Co. #2	1915 American-LaFrance cities service ladder truck
8 Fairfield Avenue	
Engine Co. #16	1923 Mack 500 gpm pumper
636 Blue Hills Avenue	

* Two piece company with two pumpers responding to alarms as one unit

Stations and apparatus of the Hartford Fire Department in 1937 are listed here. (Courtesy of the Connecticut Fire Museum Archives.)

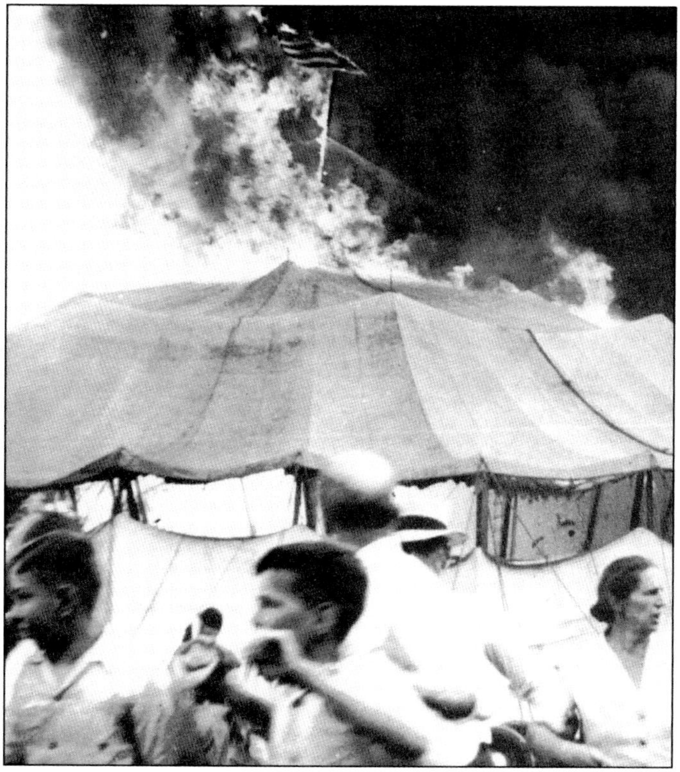

On July 6, 1944, the Barnum and Bailey Circus tent erupted into flame during an afternoon show. At 2:44 p.m., box 82, at Clark and Westland Streets, was pulled for the fire. Engines No. 7, No. 2, and No. 16, and Trucks No. 3 and No. 4 responded. A crowd of over 7,000, including many children, was on hand that day for the 2:00 p.m. performance. By the time the fire was over, 168 people were dead and hundreds severely injured. (Courtesy of the Connecticut Fire Museum Archives.)

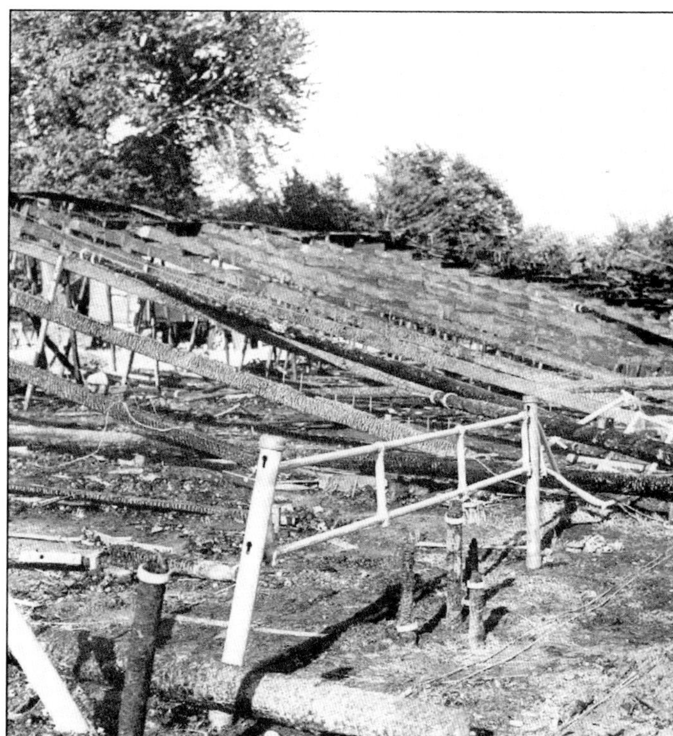

Remains of the bleacher sets give an indication of the intensity of the fire that day. New fire regulations and ordinances for outdoor tent shows were quickly enacted in the United States as a result of this tragic event. (Courtesy of the Connecticut Fire Museum Archives.)

Four

Hartford in the Modern Era

Starting in 1948, the Hartford Fire Department saw vast changes in departmental operations. These improvements, suggested by an independent research institute, included institution of the 56-hour workweek with a three platoon system, which, by reducing the average workweek from 77 hours, was an important improvement in the working conditions of the firefighters. Also, as a result of this study, the department made the first major communication improvement since 1919 by having FM radios installed in all apparatus and other department vehicles in 1948. This allowed the dispatchers to be in contact with the apparatus at all times while out of the station. Three new pieces of equipment from American LaFrance entered service: a 750-gallon-per-minute pumper for Engine No. 2, a crash truck for Engine No. 6, and an 85-foot aerial ladder for Ladder No. 2. On October 4, after successfully competing in an examination, six of six African American firefighters were hired, the first black firefighters since 1914 when William H. Jacklyn retired. Between them, they served over 170 years.

More equipment arrived in 1949 in the form of two pumpers and one aerial ladder from American LaFrance, a pumper from Mack, and the first Seagrave apparatus in the city—a 65-foot aerial ladder. These pieces were placed at various companies in the city. Again in 1951, the city purchased three more pieces: one pumper and two aerial ladders from American LaFrance. The year 1953 saw Hartford win the U.S. Chamber of Commerce Grand Award for fire prevention for a third year. This award was presented annually to the city judged best in the nation in the Inner Chamber Fire Waste Contest. Over the next few years, the city bought five more pumpers from American LaFrance. By then, all the equipment from the 1920s and 1930s was either disposed of at auction or put on reserve status. Interestingly enough, one 1928 and three 1930 American LaFrance pumpers remained on reserve status until the mid-1970s.

In 1956, Hartford saw multiple-alarm fires ravage two historic churches. The first occurred on December 29 at St. Patrick's Church, at the corner of Ann and Church Streets, and heavily damaged the interior of the church, including shattering many of the stained-glass windows due to the heat and destroying most of the roof, in a three-alarm fire. For the third time in its history, the church was ravaged by fire and later rebuilt and stands today. Less than 33 hours later, at 7:00 a.m. on December 30, a fire broke out in the basement of St. Joseph's Cathedral on Farmington Avenue. Ironically, the deputy fire chief was attending the mass, which was being said by Father O'Neill, the chaplain of the fire department, when the fire was discovered. Before the fire was under control, three alarms were struck, a 35-foot-wide ball of flame burst through the huge window in the front of the building, and the fire broke through the roof of the building. By noon, the building was considered a total loss. Many priceless historic religious artifacts were destroyed. Damages were estimated at over $5 million. This was the largest single-structure loss in the history of the Hartford Fire Department.

In 1958, the first major renovation in the city in modern history began with the construction of Constitution Plaza. This called for the demolition of Engine Company No. 3's quarters on Market Street. Engine No. 3 was relocated to Engine No. 4's quarters on Pearl Street. In 1960, three 1,000-gallon-per-minute pumpers were purchased from Mack Trucks for Engines No. 1, No. 3, and No. 5, these were some of the last open-cab apparatus that Hartford would purchase.

December 8, 1961, was a tragic day for Hartford, with a third-alarm fire at Hartford Hospital, which originated in a rubbish shute and extended down the hallway, killing 16 people. Even though the ladder trucks in the city could only reach the eighth floor, many lives were saved by the firefighters on the top of the aerial ladder instructing those on the ninth floor how to prevent the fire from coming into the patient rooms.

In 1962, Hartford raised money with a municipal bond and contracted the firm of Ebbets, Frid and Prentice to build four new stations. Engine Companies No. 3 and No. 7 moved to 181 Clark Street. Engines No. 11 and No. 12 and Ladder No. 5 moved to 150 Sisson Avenue, and Engine No. 8 had new station built on the same site as the 1890s building.

The late 1960s saw racial unrest throughout the country, and Hartford was no exception to this. Numerous buildings were set on fire, and firefighters were needed for extra shifts, sometimes working up to 22 hours a day. Long hours combined with low pay led to further adjustments in the hours that firefighters worked and how they were recruited. Hours were reduced to 42 hours per week, and a nationwide recruitment was begun. This resulted in the recruitment of experienced New York City firefighters who then relocated to Hartford and joined the fire department.

Changes in the ethnic makeup of the city and the changing times of the country also created changes in the Hartford Fire Department. The increasing minority population in the city lead to a recruit class of 50 firefighters in June 1980, 25 of whom were Hispanic, 13 of whom were black, and 23 of whom were white. It was not until 1982 that Hartford saw its first two female firefighters. Modernization of equipment also occurred during this period. The first tower ladders were purchased. Engine companies were reorganized and the first tactical units introduced to the city. These companies were equipped and trained for rescue. Initially these units used converted vintage 1950s pumpers until two Ford Ranger rescue trucks were delivered.

Over the years, Hartford continued to purchase American LaFrance apparatus until the 1980s, when Grumman, Ferrena and Sutphen began to deliver apparatus to the city. In spite of many changes over the years, Hartford remains a Class 1 fire department, one of only 25 in the United States, providing superior quality fire and EMS protection. Operating out of 12 stations, the department consists of 11 engine companies, five ladder companies, and one tactical unit headed today by Chief Charles Teale Jr.

Henry G. Thomas served as chief of the department from 1946 to 1959. It was during his term as chief that the 56-hour week was established and the installation of two-way FM radios in all apparatus and other department vehicles occurred. World War II had seen huge advances in FM radio technology, and equipping Hartford's apparatus with radios provided greater mobility and flexibility in operations. (Courtesy of Mike Gorman.)

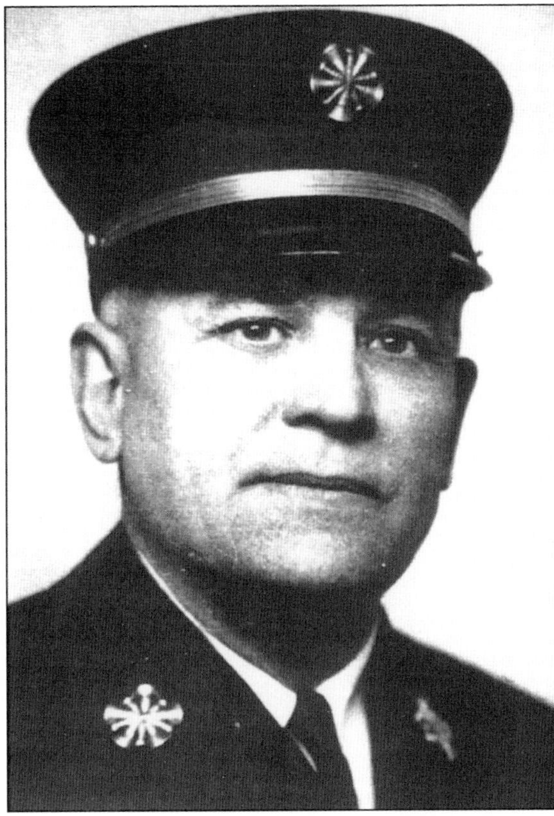

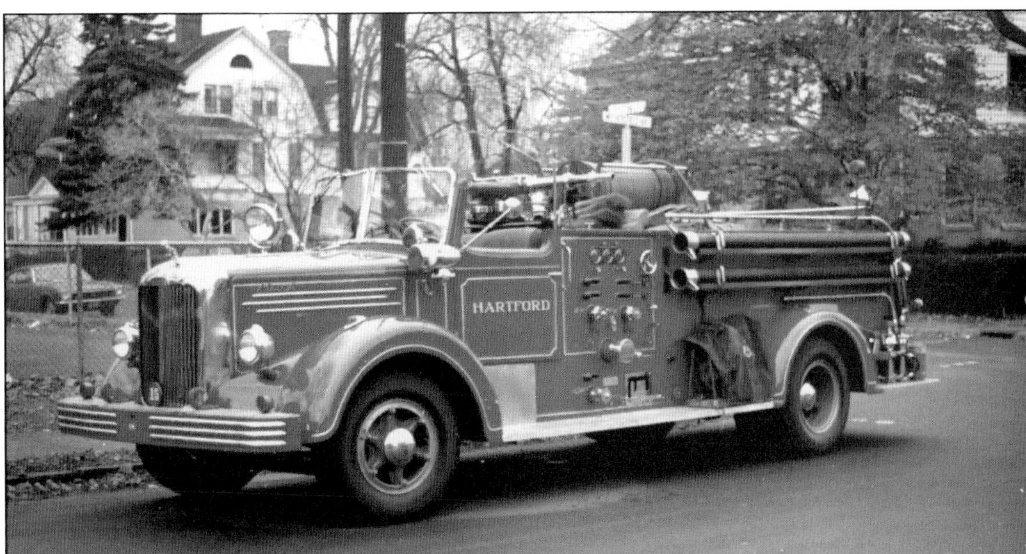

Engine No. 15's 1949 Mack type 95 pumper is shown in this November 1969 view. This was a Mack L-Model, introduced in 1940 and produced until the mid-1950s. Open cabs would soon become obsolete after the mid-1960s because of violent civil unrest and the need for firefighters to be protected from bricks and bottles thrown by rioters. By the 1980s, apparatus manufacturers were producing enclosed-cab vehicles almost exclusively. (Courtesy of Mike Gorman.)

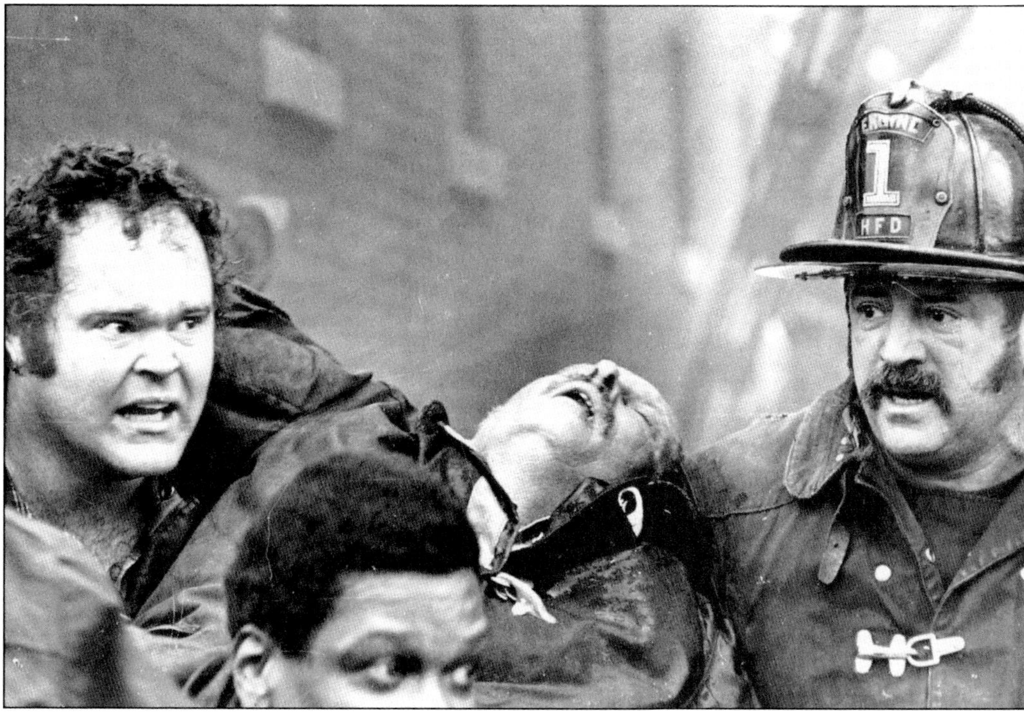

Heroism is seen in action in this photograph. An injured firefighter is being carried by his comrades during a three-alarm fire on Park Street. (Courtesy of Mike Gorman.)

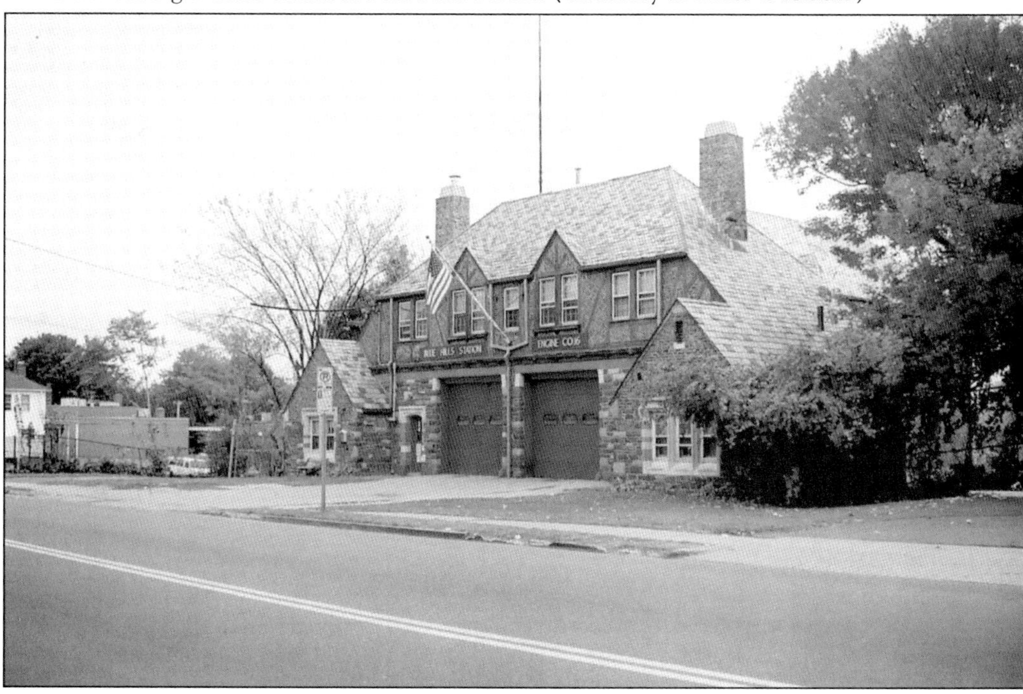

Engine No. 16's bungalow fire station on Blue Hills Avenue is seen here in 2006. Bungalow stations were built mostly in residential areas to blend in with the houses. (Courtesy of the Connecticut Fire Museum Archives.)

This is a list of stations and apparatus of the Hartford Fire Department in 1977. (Courtesy of the Connecticut Fire Museum Archives.)

1977
Stations and Apparatus

Engine Co. #1*	1953 American-LaFrance 750 gpm pumper 1961 Mack 1000 gpm pumper
Ladder Co. #6	1955 American-LaFrance 100' tiller
Lighting Wagon #1	1956 International light wagon
197 Main Street	
Engine Co. #2*	1967 Mack 1000 gpm pumper 1965 Mack 1000 gpm pumper
Ladder Co. #3	1963 Seagrave 100' aerial ladder
1515 Main Street	
Engine Co. #3*	1952 American-LaFrance 1000 gpm pumper 1960 Mack 1000 gpm pumper
Engine Co. #7	1951 American-LaFrance 1000 gpm pumper
182 Clark Street	
Engine Co. #4*	1976 American-LaFrance 1000 gpm pumper 1961 Mack 1000 gpm pumper
Ladder Co. #1	1977 American-LaFrance 100' aerial
275 Pearl Street	
Engine Co. #5	1971 American-LaFrance 1000 gpm pumper
129 Sigourney Street	
Engine Co. #6*	1954 American-LaFrance 1000 gpm pumper 1949 Mack 750 gpm pumper
Crash Truck	1948 American-LaFrance 750 gpm pumper and foam truck
98 Huyshope Avenue	
Engine Co. #8*	1971 American-LaFrance 1000 gpm pumper 1958 Mack 750 gpm pumper
721 Park Street	
Engine Co. #9	1971 American-LaFrance 1000 gpm pumper
655 New Britain Avenue	
Engine Co. #10	1971 American-LaFrance 1000 gpm pumper
520 Franklin Avenue	
Engine Co. #11	1951 American-LaFrance 1000 gpm pumper
Ladder Co. #5	1968 Maxim 100' aerial
150 Sisson Avenue	
Engine Co. #14	1971 American-LaFrance 1000 gpm pumper
Ladder Co. #4	1971 American-LaFrance 100' aerial
25 Blue Hills Avenue	
Engine Co. #15	1953 American-LaFrance 750 gpm pumper
Ladder Co. #2	1951 American-LaFrance 100' aerial
8 Fairfield Avenue	
Engine Co. #16	1971 American-LaFrance 1000 gpm pumper
636 Blue Hills Avenue	

* Two piece company with two pumpers responding to alarms as one unit

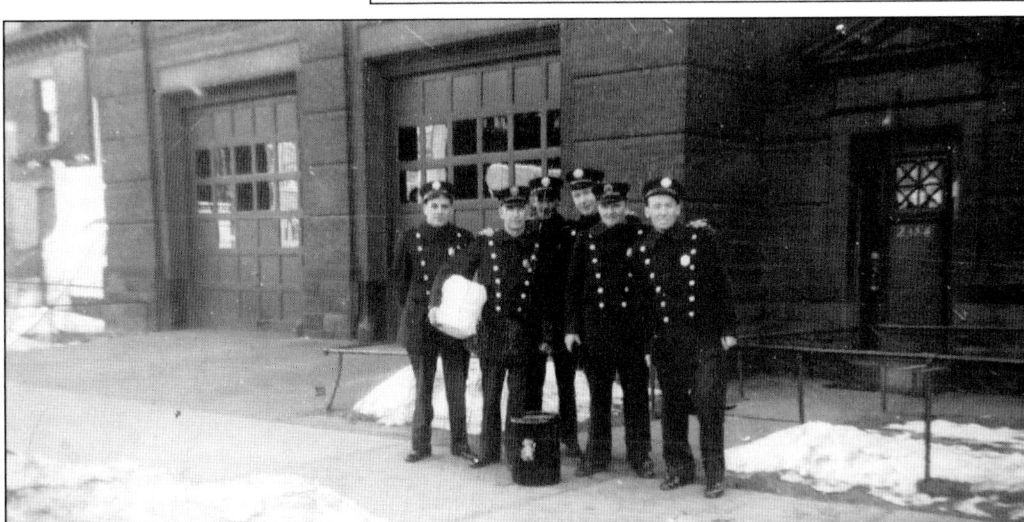

The firefighters at Engine Company No. 12 are shown in this February 1947 view. From left to right are E. Mulbarbey, Dan Welch, Mike Coughlin, Tom Moriarty, George McGahn, and Lt. Joe Winslow. (Courtesy of Mike Gorman.)

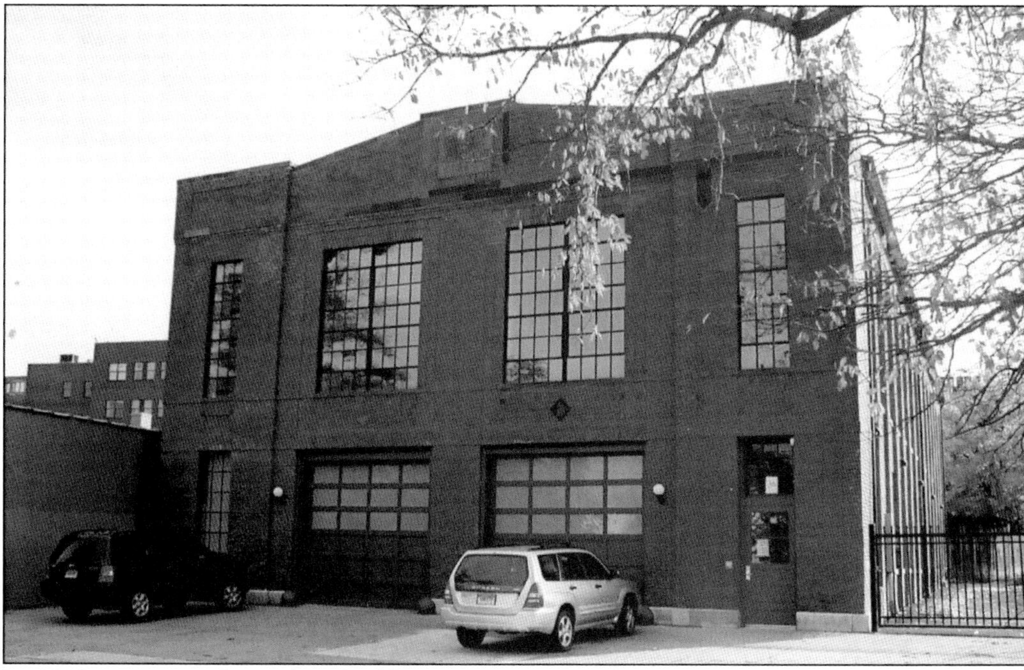

In 1920, a new machine shop building was constructed at 36 John Street; prior to that, the mechanical repair department operated out of Engine No. 7's quarters. (Courtesy of the Connecticut Fire Museum Archives.)

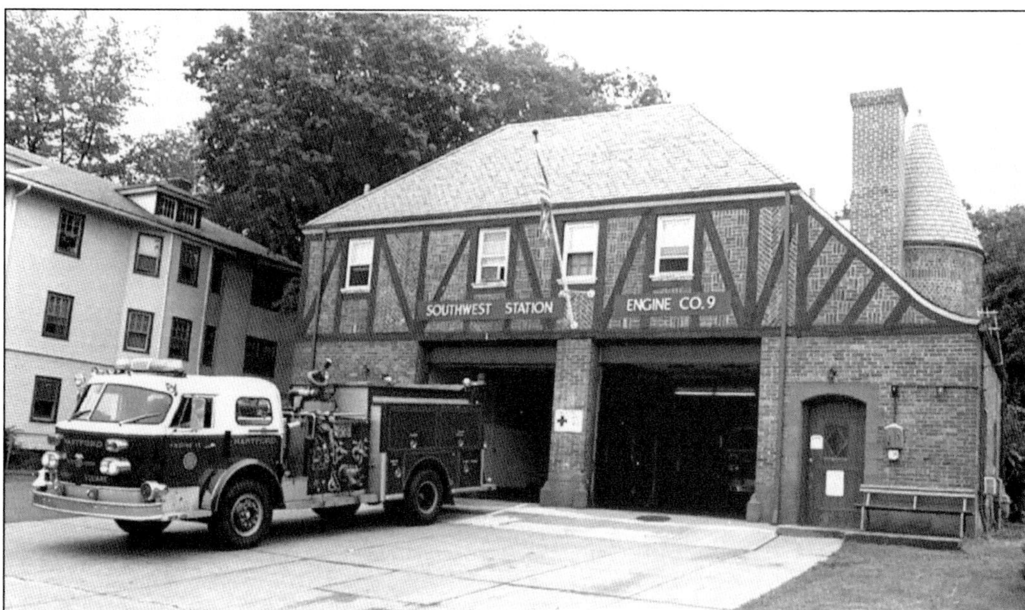

The other bungalow fire station in Hartford was Engine Company No. 9's quarters at 655 New Britain Avenue. Engine Company No. 9 was originally organized in 1900 at 43 Pearl Street as Chemical No. 1, in 1911 was renamed Squad A, and responded to all box alarms in the city. In 1928, the squad was disbanded, and Engine Company No. 9 reorganized again in 1931 when this fire station was opened. (Courtesy of Mike Gorman.)

This is a list of the chiefs of the Hartford Fire Department from 1864 to 2006. (Courtesy of the Connecticut Fire Museum Archives.)

Chiefs of the Hartford Fire Department

Henry P. Seymour	1864-1868
Henry J Eaton	1868-1903
Louis Krug	1903-1912
Augustus Loomis	1912-1913
John C. Moran	1913-1937
Michael T. Kenna	1937-1941
Thomas J. Skelley	1942-1942
John C. King	1942-1946
Henry G. Thomas	1946-1959
Thomas F. Lee	1959-1964
Edward M. Curtin	1965-1972
Edward F. Fennelly	1972-1979
Charles Gallon*	1979-1980
John B. Stewart, Jr.	1980-1992
Nelson K. Carter, Sr.	1992-1993
John Vendetta*	1994-1995
Charles A. Teale, Sr.*	1995
Robert E. Dobson	1995-2000
Charles A. Teale, Sr.	2000-

* Acting Chief

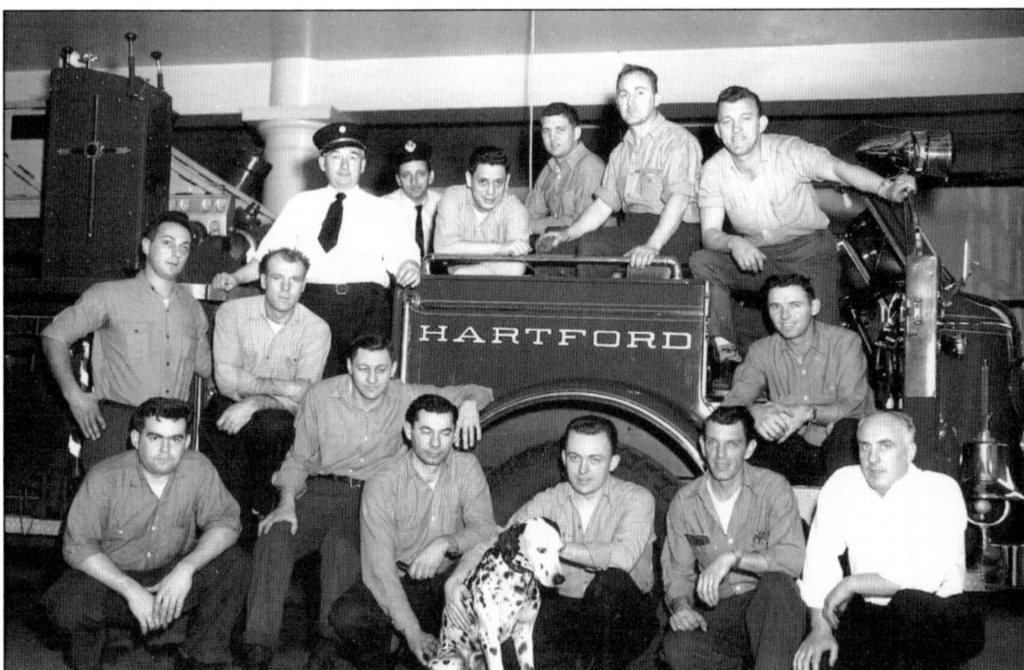

The firefighters of Engine No. 4 and Ladder No. 1 are shown in this c. 1960 photograph. (Courtesy of Mike Gorman.)

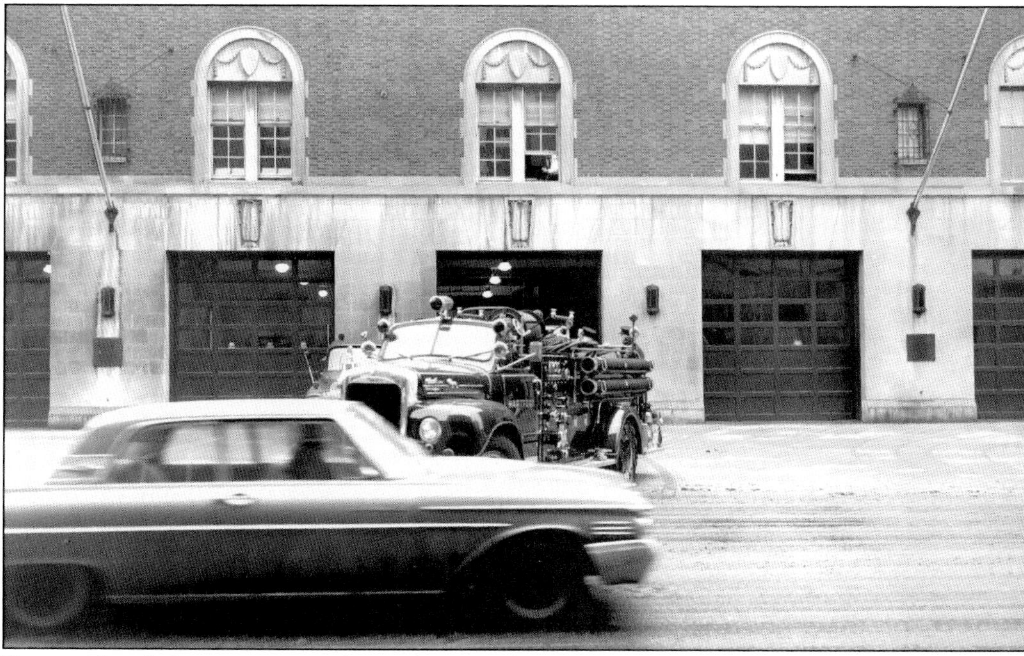

Engine No. 4's hose wagon is shown leaving the station in this 1960s view. Hartford had six companies that operated two pumpers that responded to alarms as a unit. The first truck was called the "hose wagon," the following unit the "engine." Engine No. 4 was the first unit in the city to use this concept in 1934. (Courtesy of Mike Gorman.)

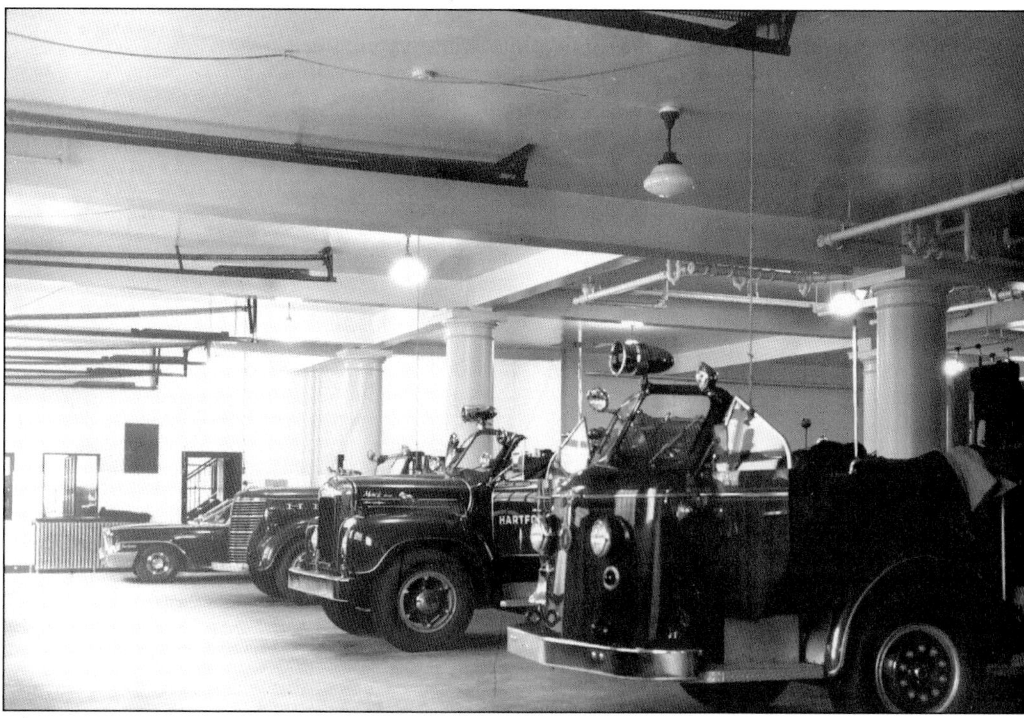

This is a 1960s view of the interior of fire headquarters on Pearl Street, with Ladder No. 1 and Engine No. 4's two pieces. (Courtesy of Mike Gorman.)

Seen here is a demonstration of the new 85-foot aerial ladder, built in the Hartford Fire Department's own machine shop. The photograph was taken in front of Engine No. 1's quarters on Main Street. (Courtesy of Mike Gorman.)

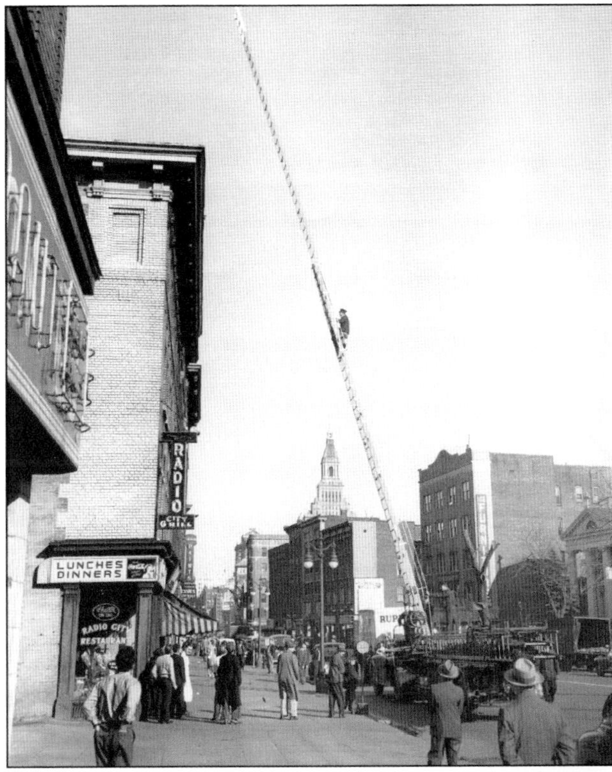

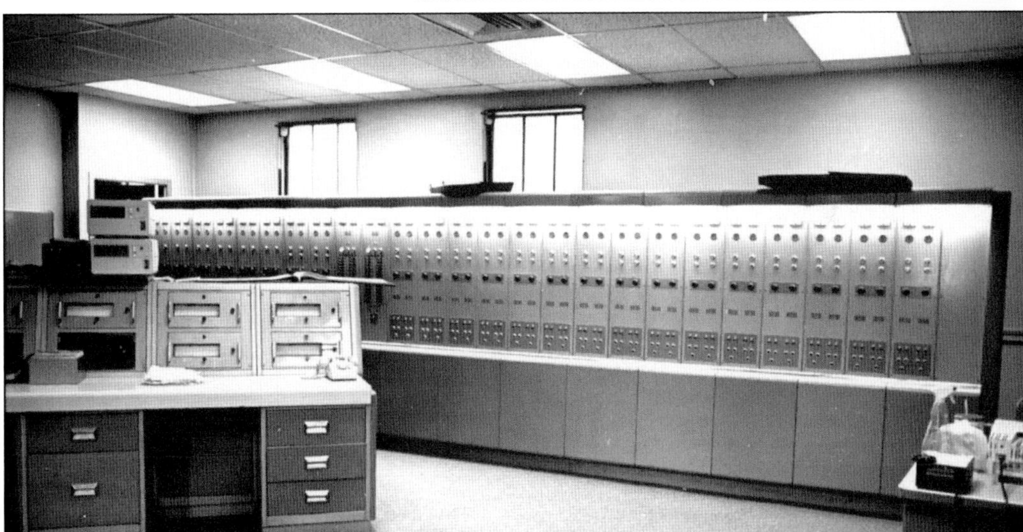

The Fire Alarm Center was located on the upper floors of Engine No. 4's quarters. In this 1980s view, the Gamewell equipment seen was used to operate the system. Hartford had 750 fireboxes located around the city at intersections and inside major buildings. The 25 circuits were controlled from the panel along the wall. The office was staffed with firefighters who worked eight-hour shifts. In 1987, all dispatching for the fire department was transferred to civilian dispatchers, and in May 1989, the Fire Alarm Center was combined with the police department into a common dispatcher center inside the police station on Jennings Road. (Courtesy of Mike Gorman.)

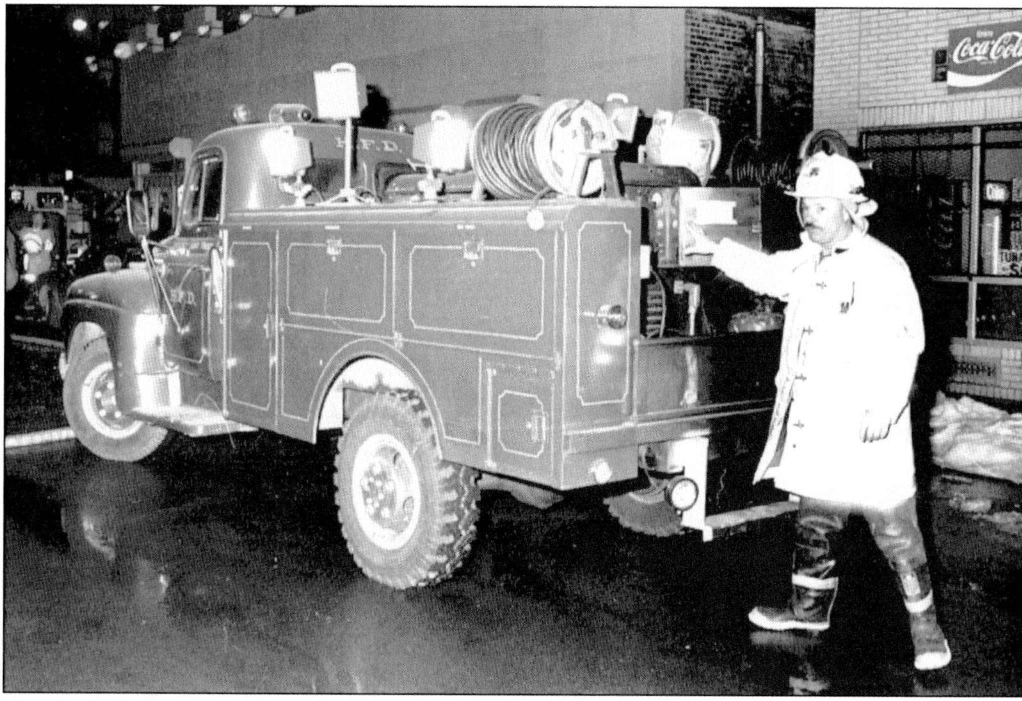

Hartford's lighting truck is shown at a fire on Guilford Street, a rare sight as this unit only responded to multiple alarms during the night. Box 146 was struck for this factory fire that was destroyed after three days of burning. This was one of Hartford's longest burning fires. (Courtesy of Mike Gorman.)

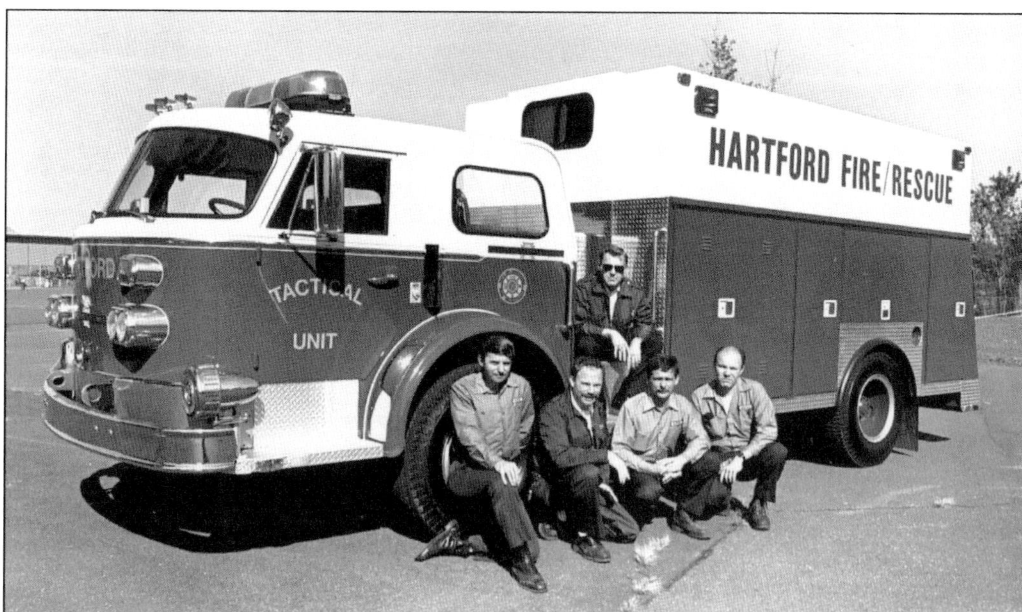

A spare tactical unit was built by Hartford's machine shop force from a 1971 American LaFrance pumper that was originally used by Engine No. 9. Hartford's mechanical department has been refurbishing and rebuilding fire apparatus since the 1920s. (Courtesy of Mike Gorman.)

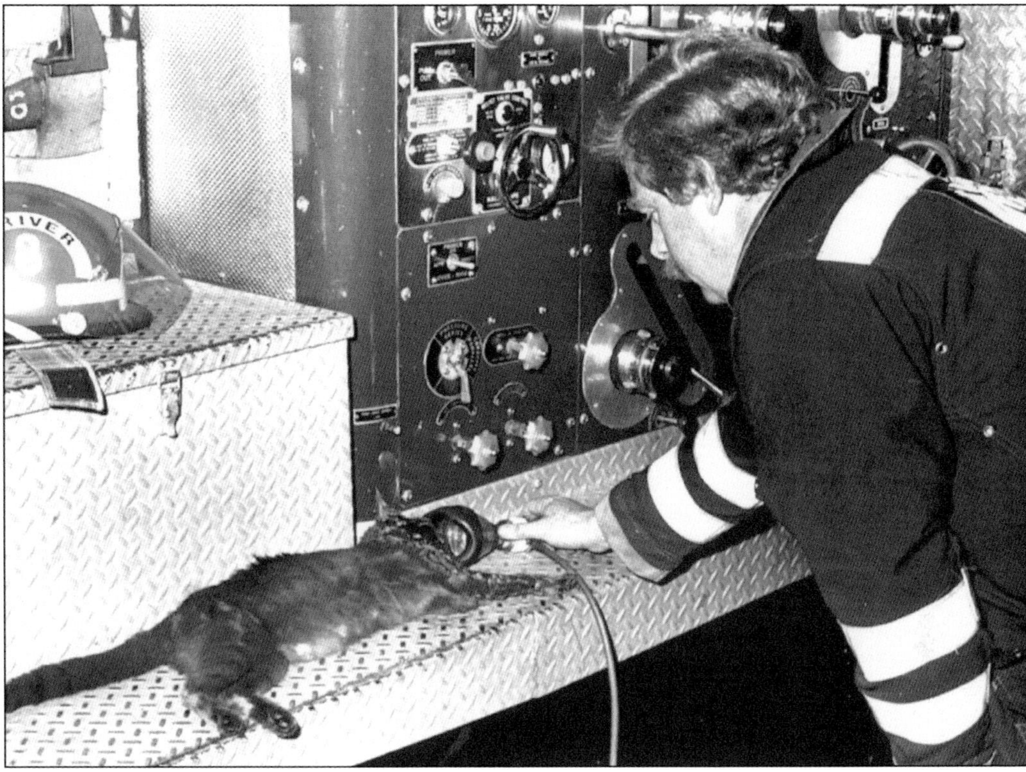

A firefighter from Engine No. 8 is trying to revive a pet caught in an apartment house fire on Lincoln Street in 1987. In the horse-drawn era, almost all fire stations had one or more house cats living in quarters to alleviate the rodent problems associated with the storage of grain and oats for the horses. (Courtesy of Mike Gorman.)

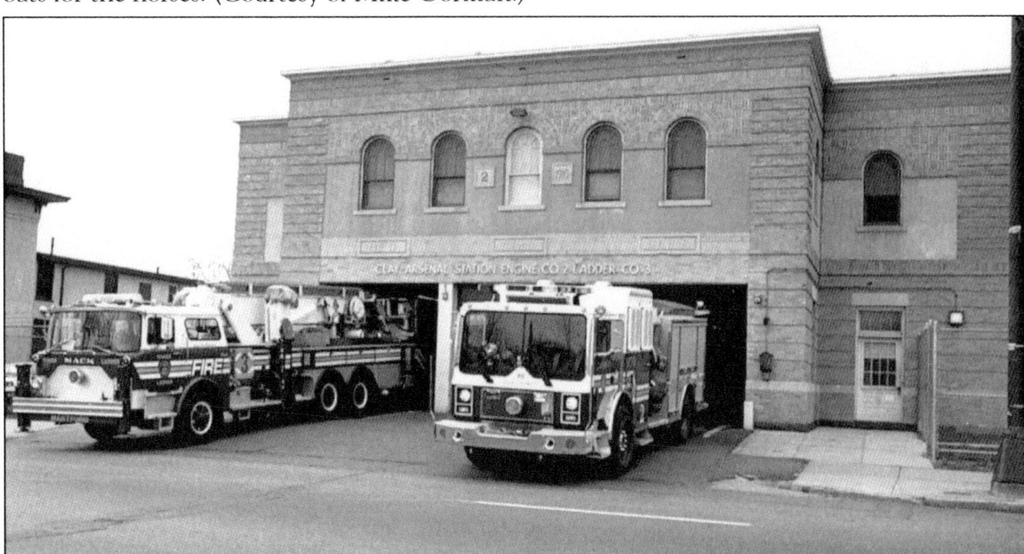

Engine No. 2's quarters are seen here in 1996. Compare this to the picture on the bottom of page 25. (Courtesy of Mike Gorman.)

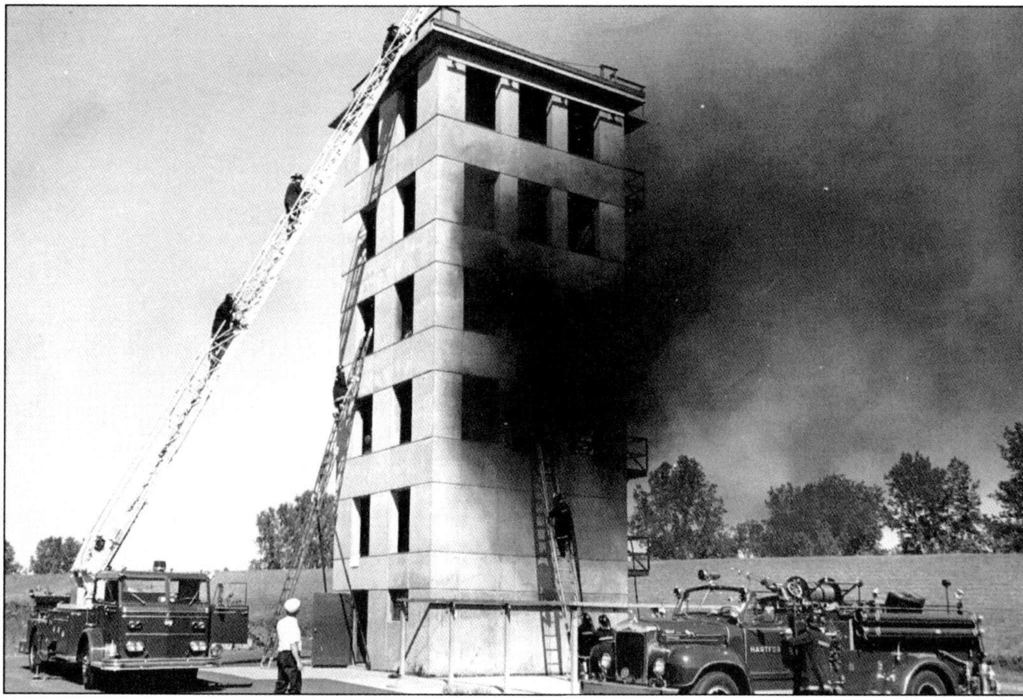

Hartford's training tower is seen here in 1970, with Engine No. 1's Mack pumper and Ladder No. 5's 1968 Maxim 100-foot aerial ladder. This was Hartford's only Maxim apparatus. Maxim equipment was very popular with other departments such as West Hartford, which had all Maxim apparatus during the 1950s. (Courtesy of Mike Gorman, photograph by Edward Lescoe.)

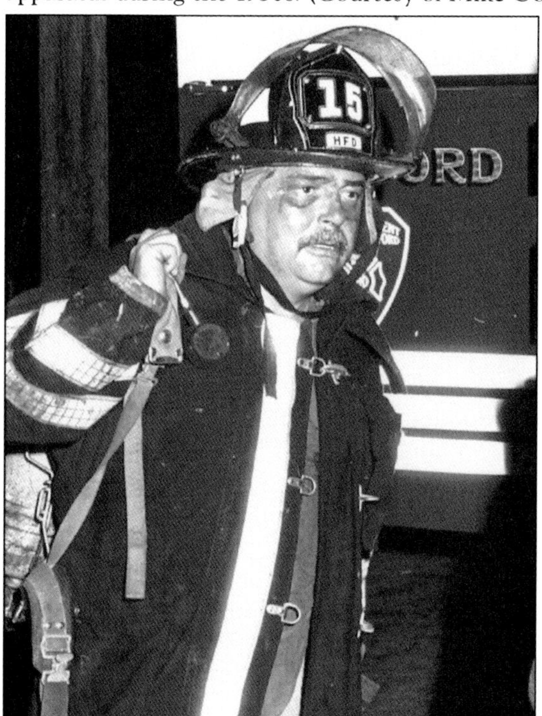

Capt. Mike Peters of Engine Company No. 15 served as mayor of Hartford after retiring from the department. He served as mayor from 1993 to 2000. (Courtesy of Mike Gorman.)

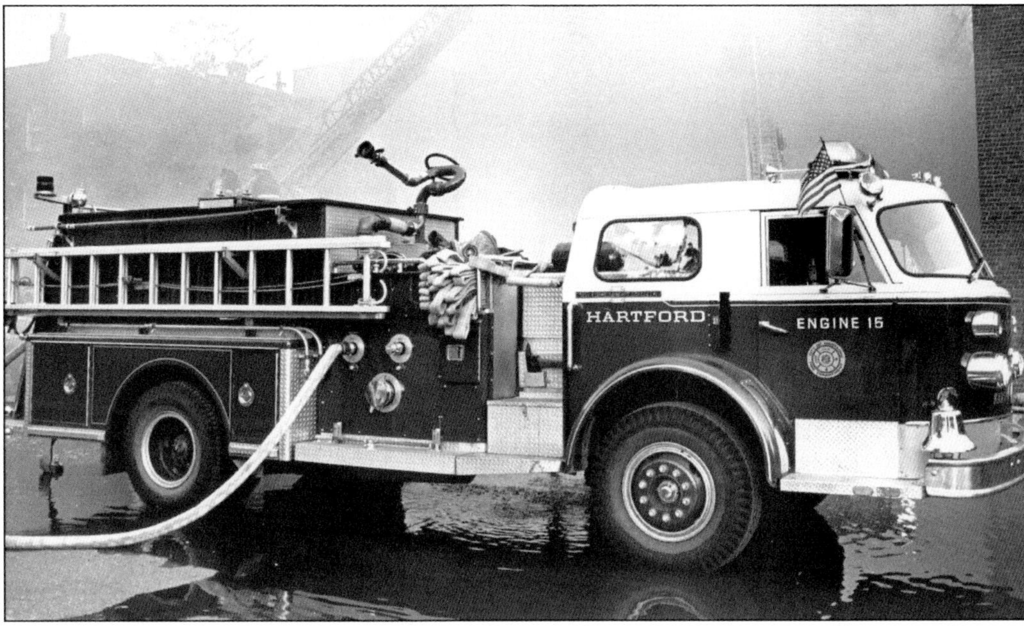

Engine No. 15's American LaFrance pumper is seen here at a three-alarm fire on Washington Street in 1989. Compare this to the 1928 American LaFrance at the top of page 50. At one time, American LaFrance was considered the "General Motors" of fire apparatus manufacturers. American LaFrance produced the first cab-forward fire apparatus in 1939. Today the majority of custom built apparatus are constructed with the cab-forward design. (Courtesy of Mike Gorman, photograph by Joseph J. Marino.)

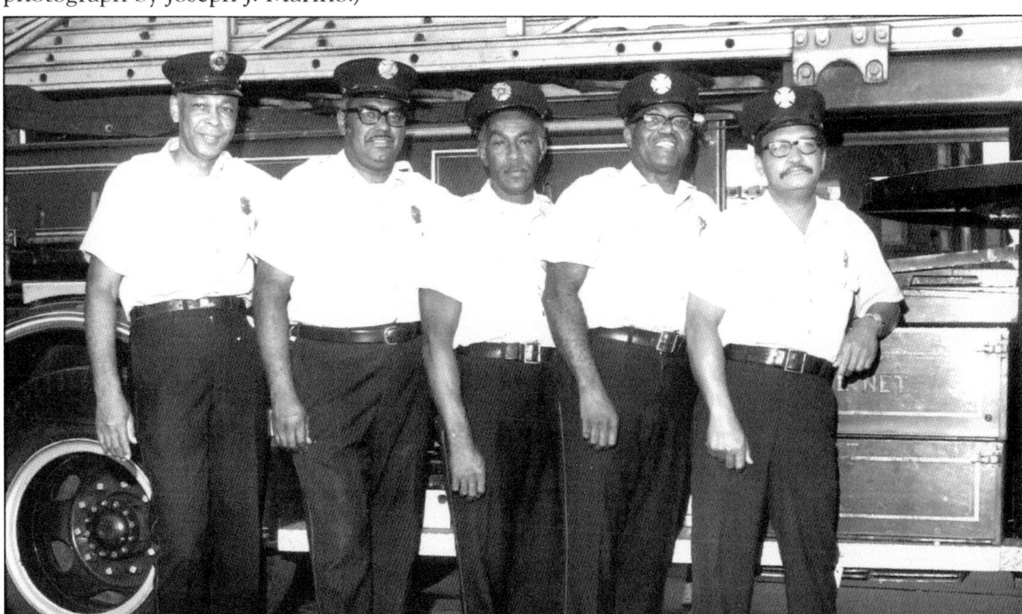

From 1914 when William H. Jacklyn retired until 1948 when six African American firefighters jointed the department, no African Americans served on the Hartford Fire Department. From left to right, George Hayes, Frank Davis, Ben Laury, James Lewis, and Harry Ashe all were hired in October 1948. (Courtesy of Mike Gorman.)

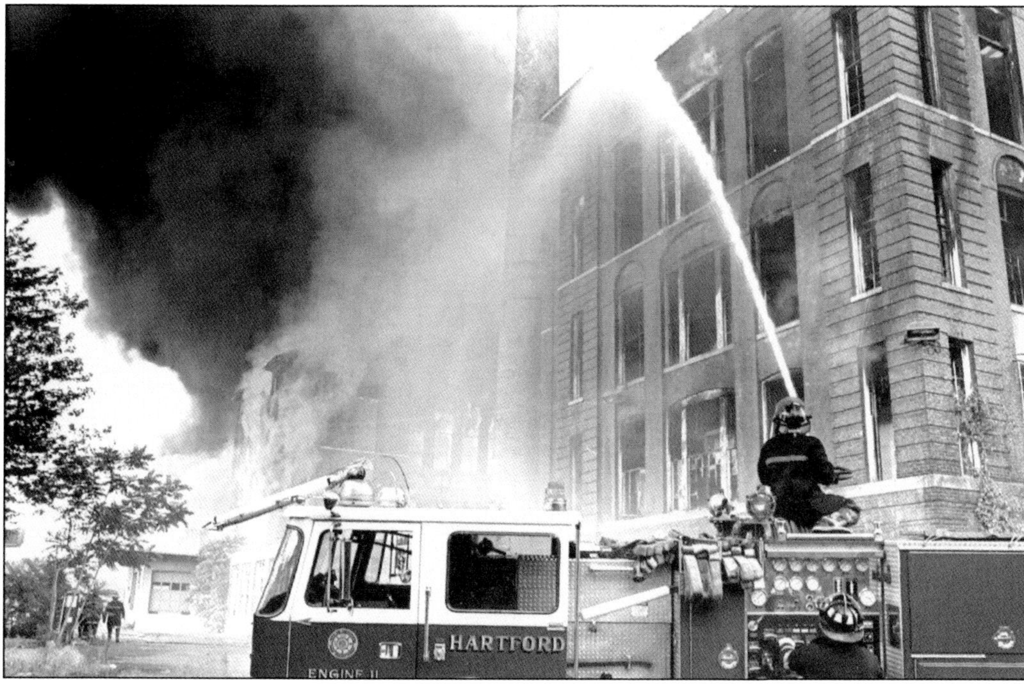

On July 12, 1992, the building formally occupied by the Royal Typewriter Company was destroyed in a three-alarm fire. In this photograph, Engine Company No. 11 is wetting down the structure. (Courtesy of Mike Gorman.)

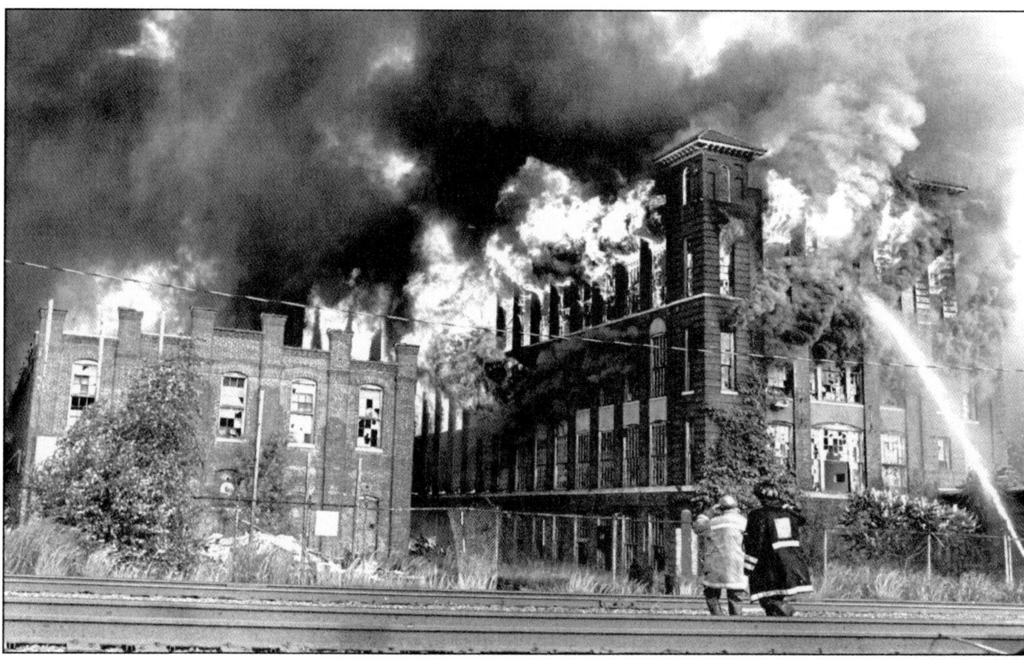

Another view of the Royal Typewriter Company fire was taken in the rear of the building. At one time, over 5,000 people worked in this building assembling typewriters. During Word War II, the plant was retooled to make machine guns for the war effort. (Courtesy of Mike Gorman, photograph by Joseph J. Marino.)

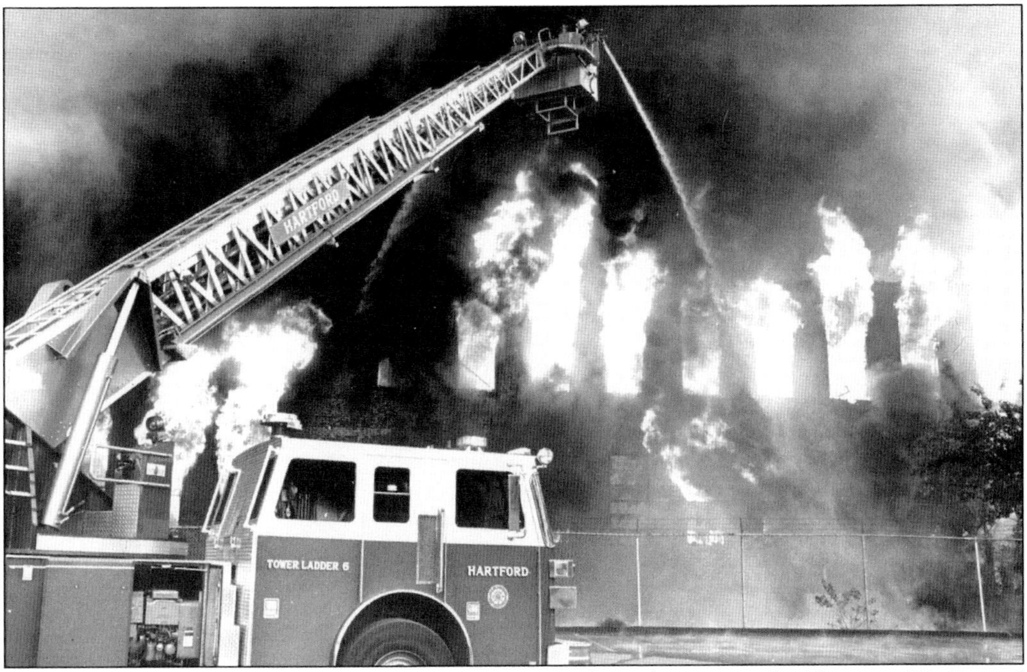

Another three-alarm factory fire occurred in 1995, when the vacant building at 612 Capital Avenue, formally occupied by Crystal Labs, was destroyed. With the advent of metal aerial ladders, the water towers developed in the horse-drawn era were rendered obsolete. (Courtesy of Mike Gorman.)

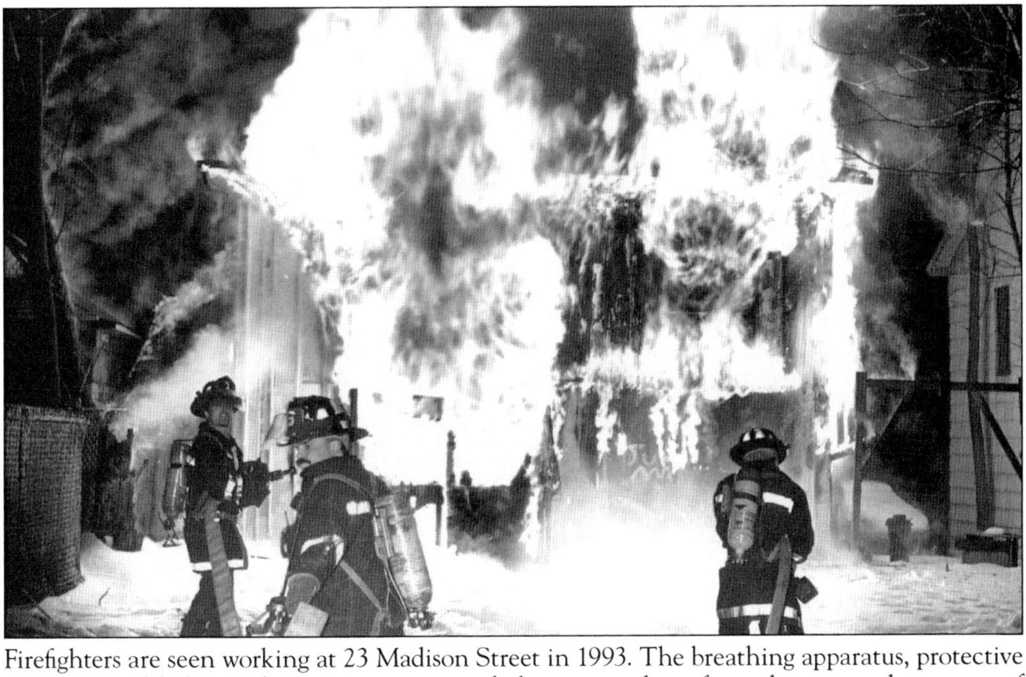
Firefighters are seen working at 23 Madison Street in 1993. The breathing apparatus, protective clothing, and helmets the men are wearing help protect them from the tremendous array of hazards that they can be exposed to at a fire. Within the past 10 years, tremendous advances have been made since the day of the rubber coats and boots. (Courtesy of Mike Gorman.)

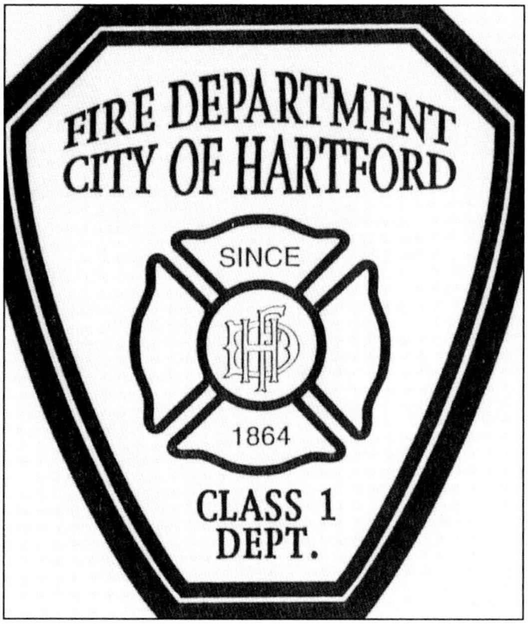

This symbol is the Hartford Fire Department insignia. (Courtesy of the Connecticut Fire Museum Archives.)

**HARTFORD FIREFIGHTERS
WHO DIED IN THE LINE OF DUTY
OR SERVICE CONNECTED ILLNESS**

Noah Risley	2/17/1881	Benjamin F. Herrick	6/17/1875
Daniel Camp	5/24/1878	Charles Harper	5/24/1875
John Parker	5/24/1878	Willie Bush	2/13/1892
Charles E. Main	2/21/1897	George Goodrich	3/28/1906
Ernest Quigley	5/17/1909	John J. Daley	12/27/1910
Anthony F. Bolan	3/4/1917	James T. Hughes	6/30/1917
Daniel Hines	10/15/1918	John J. Horan	11/2/1918
Louis E. Tucker	5/12/1919	Cornelius Quirk	12/26/1919
Oliver S. Lathrop	9/15/1922	Daniel J. Dahill	9/7/1924
Joseph O'Conner	2/16/1926	John J. McNally	3/22/1926
James W. Legeyt	11/11/1927	Robert E. Lee	4/15/1928
James O'Conner	4/21/1928	Keron Finn	12/4/1928
Michael McCormick	9/18/1929	James Costello	7/6/1930
Jeremiah Collins	1/2/1931	John Hines	10/16/1931
Phillip E. Duffy	1/3/1933	AustinE. Dugan	7/29/1933
John R. Lyons	12/12/1933	Bartholemew Rogers	5/23/1934
John J. McCarthy	10/31/1935	Edward T. Lyons	1/5/1936
Fred Shuckerow	11/1/1938	Edward J. Farrell	10/31/1939
Joseph Robertson	5/11/1940	Daniel G. Landerfin	11/14/1940
Joseph Reynolds	4/27/1941	Edward Reynolds	10/25/1941
Harold Snelgrove	3/4/1942	Kenneth P. Holley	3/16/1943
James J. O'Brian	6/22/1945	Phillip E. Sullivan	9/1/1945
James E. Slattery	12/5/1945	Daniel E. Welch	11/22/1949
William J. Scully	3/29/1951	James P. Foley	11/28/1951
Albert Hyman	10/23/1952	Marshall Slavkin	1/29/1968
Curtiss Kelley, Jr.	5/28/1970	Edward Walsh	8/16/1973
Thomas A. Fischer	9/15/1974	John Harbut	9/21/1978

This is a somber list of the Hartford firefighters who died in the line of duty. Courtesy of the Connecticut Fire Museum Archives.)

Five

East Hartford Fire Department

In 1889, the Connecticut General Assembly granted a charter to the East Hartford Fire District, giving it power to contract with the East Hartford Water Company to use fire hydrants. The district did not actually organize and contract with the East Hartford Water Company for the rental of 40 hydrants until June 1891. The fire district encompassed the most populated sections of East Hartford, the central business area up to Burnside Center. Prior to 1889, firefighting consisted of neighbors helping each other with pails. In October 1854, the Hartford Fire Department responded to a fire at the South East District School in East Hartford, providing mutual aid. This is the first record of mutual aid between these two towns. The first chief, Joseph Taft, was appointed in 1891. In 1893, the district initiated a tax of three mills on all property within its boundaries, and $3,000 was appropriated for the construction of three fire stations, Center Hose Company No. 1 on Bissell Street, Volunteer Hose No. 2 on Park Avenue on land donated by Chief Taft, and Volunteer Hose No. 3 on Hanmer Street. In 1893, citizens in the western portion of town, the Meadow Fire District, organized Aetna Hose Company No. 4, which was independent from the East Hartford District. Each company had between 35 and 40 members and operated hose carts pulled by firefighters. A bell in the center of town was used to alert the firemen; the number of strokes indicated the part of town to which the volunteers needed to respond. In 1899, the fire district purchased the East Hartford Water Company. This was just the first of many services the district provided to the residents of East Hartford in addition to fire protection. Edward O. Goodwin was the second fire chief, appointed in 1894. In 1903, the district installed 12 Gamewell fire alarm boxes along Main Street. By 1912, they expanded to 43 fireboxes, and in the 1940s, the town had a total of 85 boxes that remained in service until the early 1990s.

Being a volunteer fireman in East Hartford at the beginning of the 20th century had a bright side. Fires were few and, being a fraternal organization, there was the social life: dinners, parties, and dances sponsored by the fire company. In 1901, the department answered five alarms, by 1911, it had risen to 11, and by 1920, it was 120. A population of 6,406 in 1900 rose to 11,648 by 1920. Since organizing, the companies had gotten along with hand-drawn hose reels, except for Center Hose Company No. 1 with a horse-drawn hose and ladder wagon. In 1914, the district bought a Pope-Hartford chemical and hose car for Company No. 2. In 1916, a similar unit was purchased for Company No. 3, and in 1917, Center Hose Company received a new American LaFrance type 75 pumper and hose car.

In April 1912, the Cairns Woodworking Company on Village Street, in the Meadows Fire District, burned. Hartford sent Engines No. 2 and No. 3 to assist the East Hartford volunteers. At $225,000, this was the largest fire loss in town up to that time. Again on March 16, 1915, the Hartford Fire Department responded to East Hartford to provide mutual aid when the high school burned. East Hartford's volunteers refused the offer to connect their hoses to Hartford's steamer. The fire district sent a formal apology to Hartford Fire Department's Chief John C. Moran having realized that the Hartford Fire Department had been generous in its offer of mutual aid.

In 1921, Hose Company No. 2 on Park Avenue received a new REO chemical and hose car. The "modernization" of the fire district was completed in 1923, when Hose Company No. 3 received a new Mack AB pumper and hose car. In 1916, master mechanic Frank Meunier became the first full-time fireman. In 1921, he succeeded Edward Bragg as fire chief, a post he would hold until 1945. In 1913, the New Haven Railroad engine house on Park Avenue began to transmit fire alarms using their steam whistle. The alarms were called into the engine house by the telephone operator in the town hall, and the number of whistle blasts indicated the nearest firebox. In 1919, the fire alarm panel was moved to Center Hose Company on Bissell Street and used an air-operated horn, a diaphone, to alert the callmen. The fire district now had two full-time firefighters on duty, and the New Haven Railroad had not been properly maintaining the equipment. In 1923, Aetna Hose Company No. 4 was disbanded, and the Meadow District contracted with the East Hartford Fire District for protection. The Aetna Hose fire station was demolished in 1925 in conjunction with a road construction project on Hartford Avenue (Connecticut Boulevard today). The remaining three companies employed two paid firefighters and four part-time callmen. In 1925, the Silver Lane Fire District contracted with the East Hartford Fire District to provide fire protection for its area. Additional fireboxes were installed on Burnside Avenue, as well as installing a bell at Chief Meunier's house. In that year, the fire district adapted National Standard hose thread for all nozzles, hydrants, standpipes, hose, and apparatus instead of the previously used Bridgeport thread that had become obsolete and did not match the equipment of the companies providing mutual aide: Hartford and South Manchester.

On just one day in 1926, two fires caused over $1 million in loses. In the morning, the Stowe and Olmstead grain storehouse and the East Hartford Tobacco Growers warehouse burned with loses of $700,000. That evening, the New Haven Railroad Car and Machine Shop caught fire, causing damages of $250,000. Mutual aid was provided by Hartford, South Manchester, and Rockville. However, poor water pressure hindered the firefighters at both fires. On June 10, 1929, voters approved a new town charter that merged the East Hartford Fire District into the town of East Hartford. The first new firehouse built in East Hartford since 1893 was built in 1931 at 304 Main Street for the newly organized Engine Company No. 5, opening to protect the south end of town. In 1937, East Hartford opened a new central fire headquarters on Main Street next to the new town hall on the former Connecticut Company streetcar barn location. In 1940, Engine Company No. 2 moved into a new station at 1692 Main Street, and East Hartford received two Mack type 50 pumpers. In 1943, the last of the original 1893 stations was retired when Engine Company No. 3 moved to a new station on Burnside Avenue. In 1948, the department purchased its first ladder truck, a 75-foot aerial ladder, from American LaFrance.

Addressing the need to provide fire protection to the southeast section of town, a new station opened in 1963 at 1050 Forbes Street to house the newly organized Engine Company No. 6 to provide coverage in that area. Since Aetna Hose Company No. 4 closed in 1923, East Hartford has not had a Company No. 4. Today East Hartford is headed by Chief Michael Eremita and operates five engine companies, two ladder companies, and one rescue company from five stations with 133 career firefighters, represented by Local 1548 International Association of Firefighters. In 2005, the department had 9,487 calls.

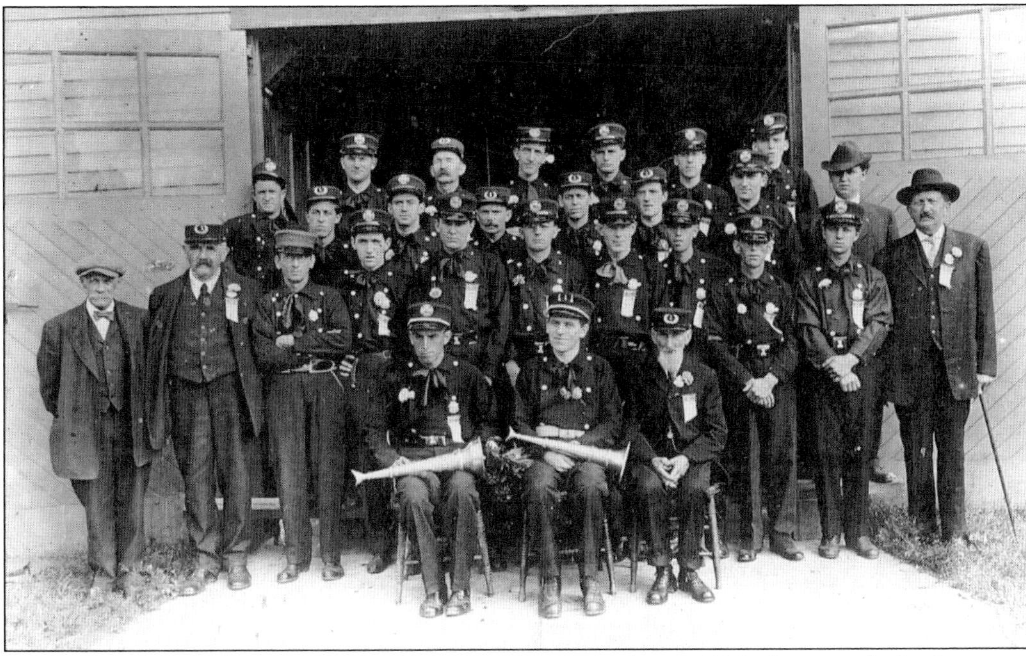

The firefighters of Center Hose Company are shown in this *c.* 1900 photograph. In this quiet, unsophisticated time before television, cell phones, and laptop computers, the dream of every boy was to be a firemen or a railroad engineer. (Courtesy of the Ken Beliveau Collection.)

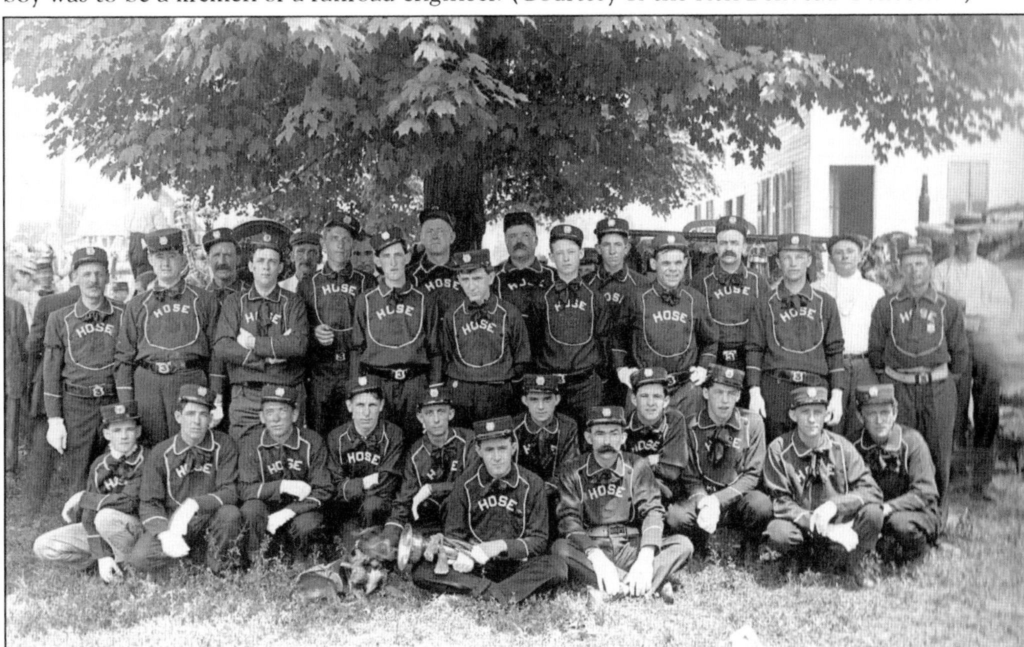

The Burnside Hose Company is seen on dress review sometime around 1900. The company's hand-drawn hose reel is in the background. In 1916, Company No. 3 received a hose and chemical truck built at a cost of $1,538.76, using a Pope-Hartford automobile. In 1923, with the arrival of a new Mack 350-gallon-per-minute pumper, the 1916 unit was retired. (Courtesy of the East Hartford Fire Department.)

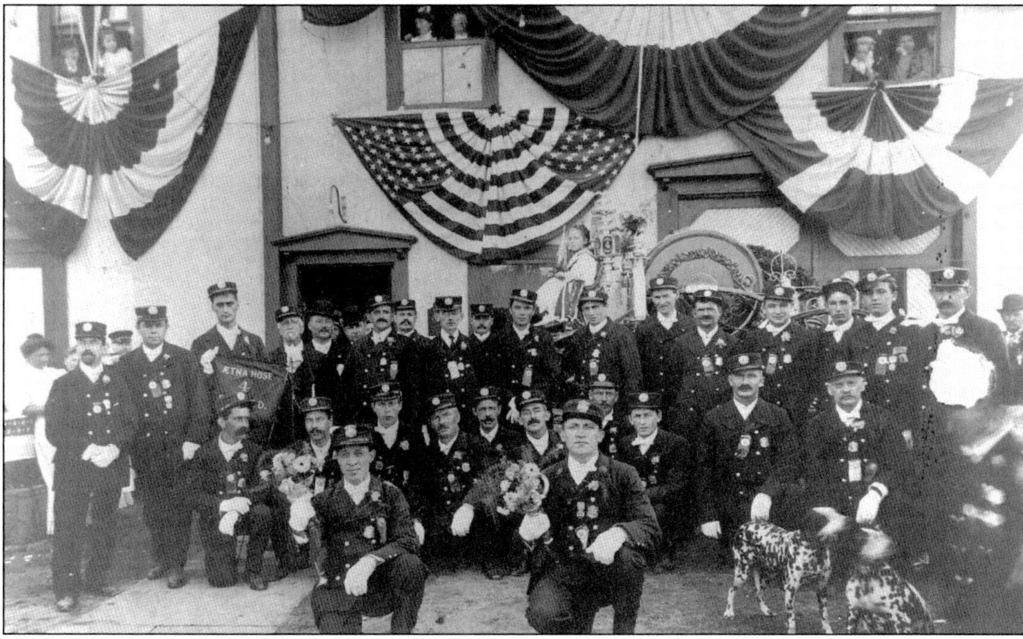

Aetna Hose Company No. 4 was not part of the East Hartford Fire District but was the fire company of the independent Meadow Fire District, organized in 1893. Their hose house was on what is now Connecticut Boulevard, a short distance from the Buckely Bridge. In 1923, the Aetna Hose was disbanded, and the fire district contracted with the East Hartford Fire District to provide fire protection. (Courtesy of the East Hartford Fire Department.)

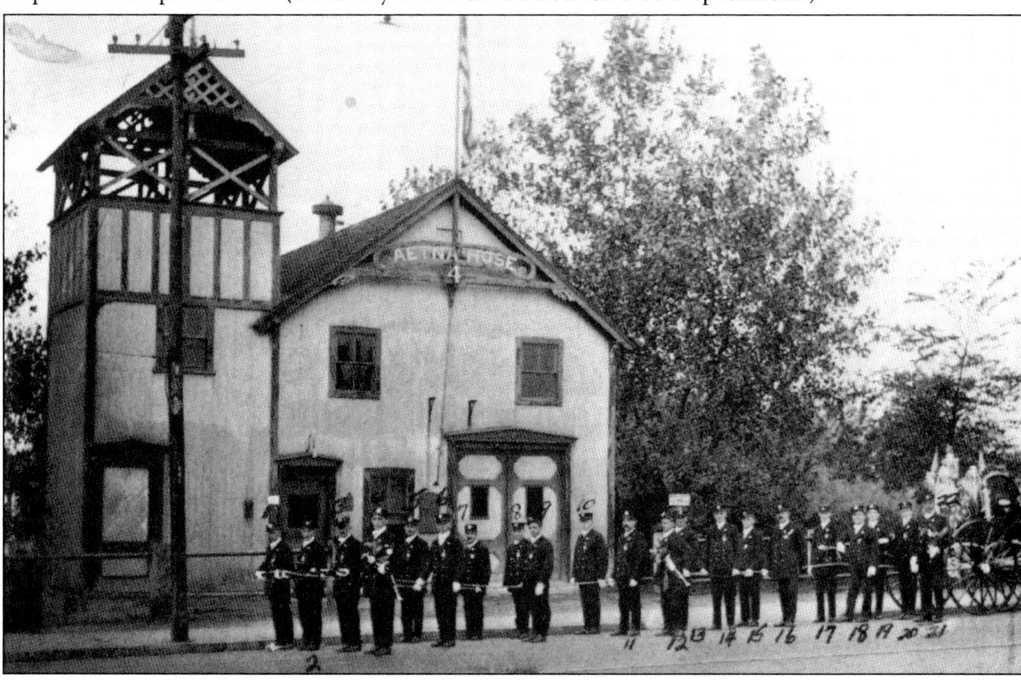

Aetna Hose Company firefighters are seen here with their hose reel. East Hartford had good water pressure from the East Hartford Water Company reservoir in Glastonbury, so a hand pumper was not required. (Courtesy of the East Hartford Fire Department.)

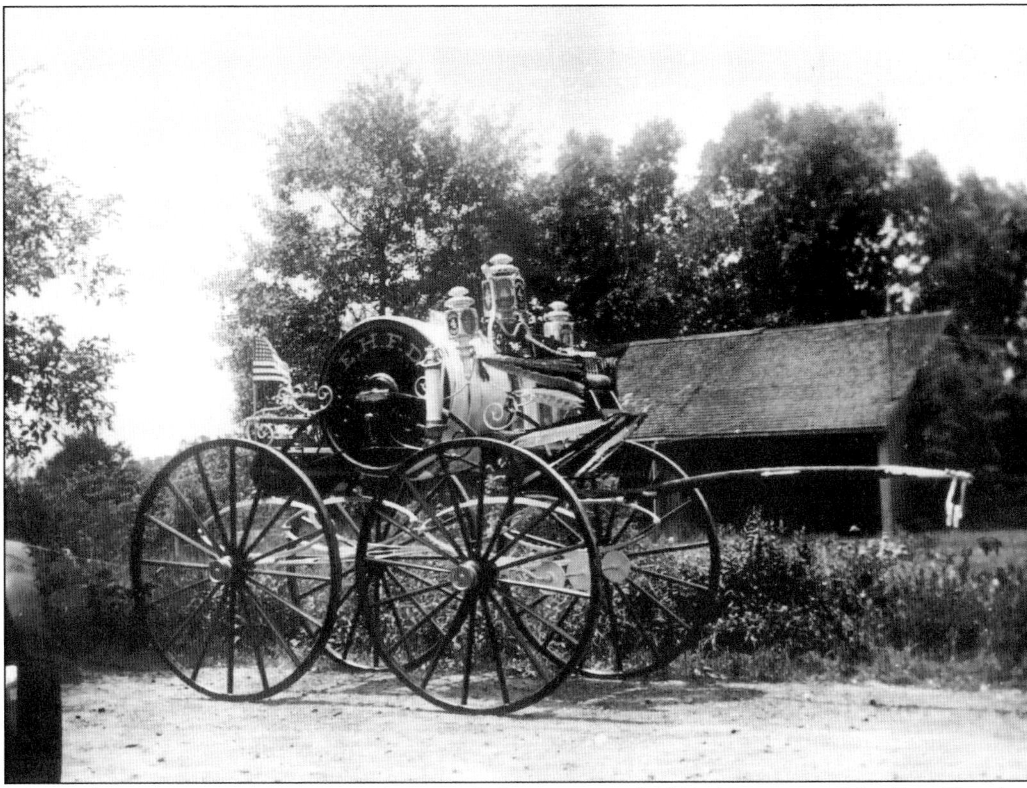

This is the Aetna Hose Company's hose reel, shown around 1910. (Courtesy of the East Hartford Fire Department.)

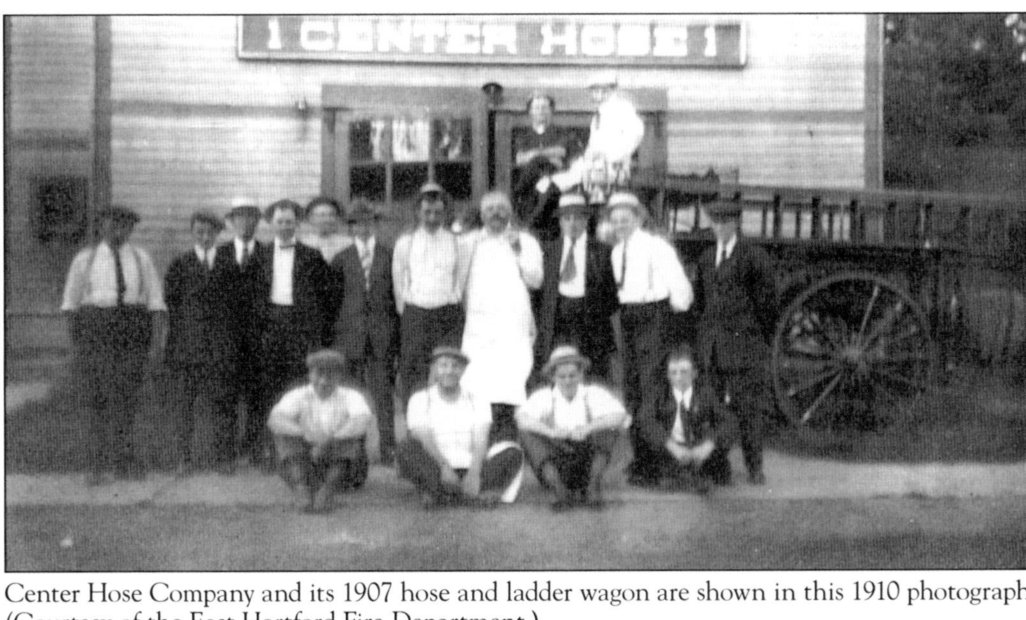

Center Hose Company and its 1907 hose and ladder wagon are shown in this 1910 photograph. (Courtesy of the East Hartford Fire Department.)

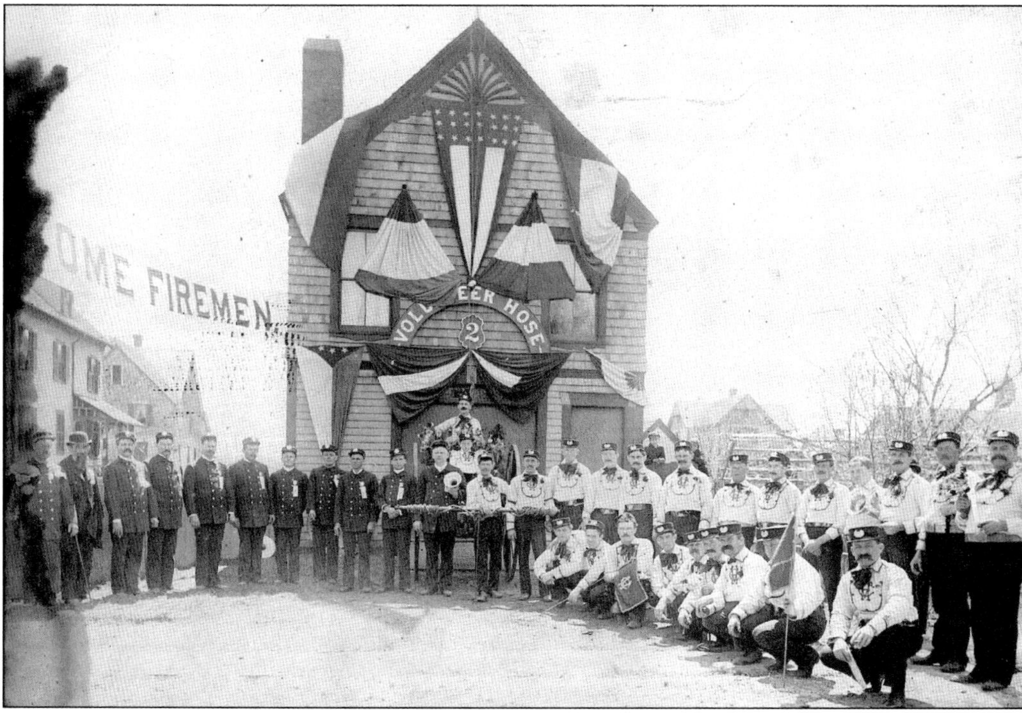

The Volunteer Hose Company No. 2 is seen here around 1900. The station was built in 1893 and was in service until 1940. (Courtesy of the East Hartford Fire Department.)

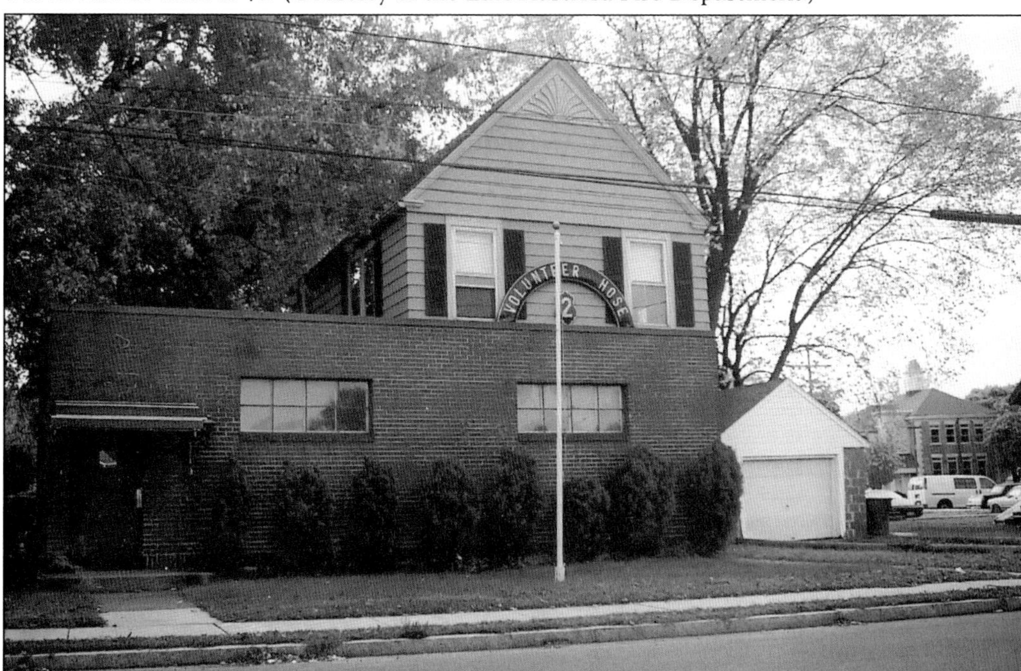

This is a 2004 view of the original station No. 2, which was built in 1893 on land donated by the chief, at that time Joseph Taft. This building was a clubhouse for volunteer firefighters until just recently. (Courtesy of the Connecticut Fire Museum Archives.)

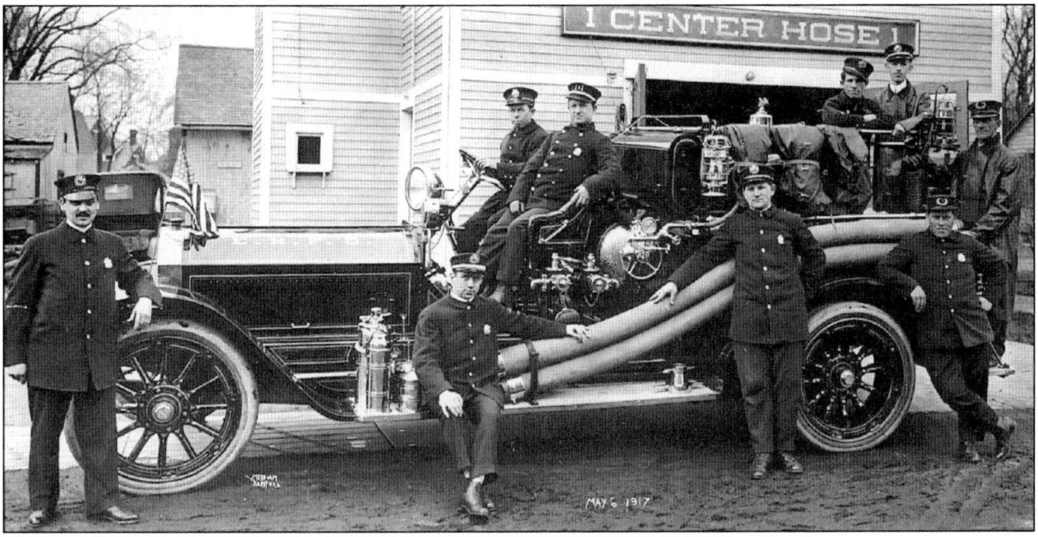

This is an image of East Hartford's 1917 American LaFrance type 75, 750-gallon-per-minute pumper, chemical and hose car shortly after delivery. In 1931, this engine was transferred to the newly organized Engine Company No. 5 in the south end of town. (Courtesy of the East Hartford Fire Department.)

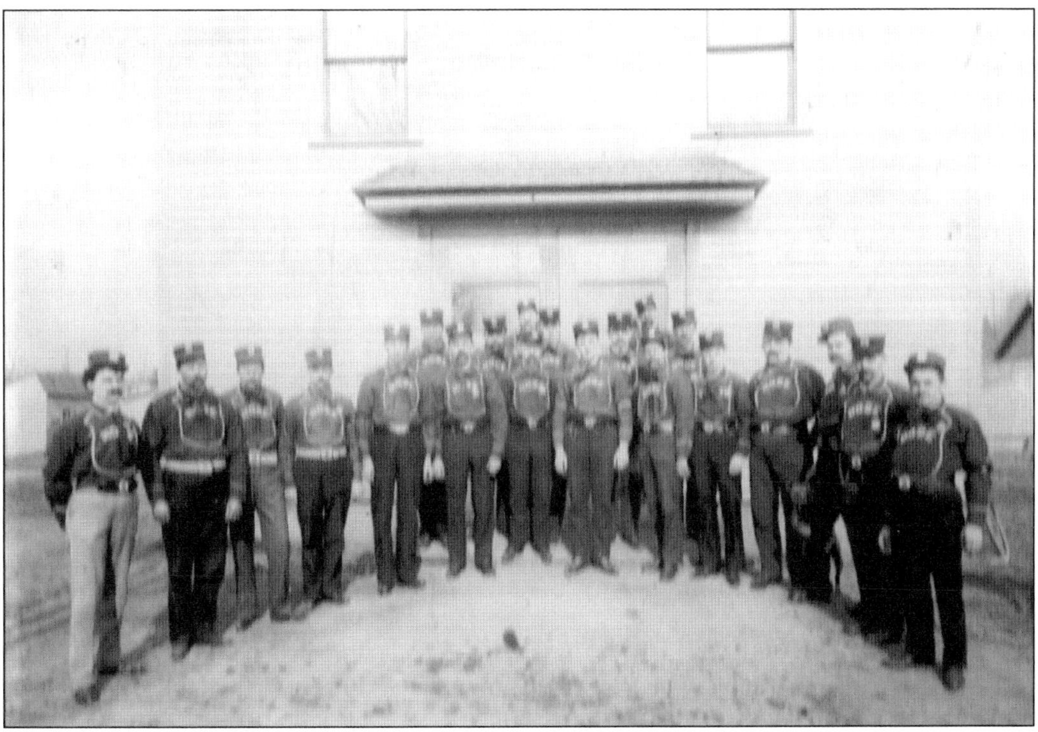

Seen here is a view of the Burnside Hose Company No. 3 quarters around 1910. (Courtesy of East Hartford Volunteer Hose Company No. 3.)

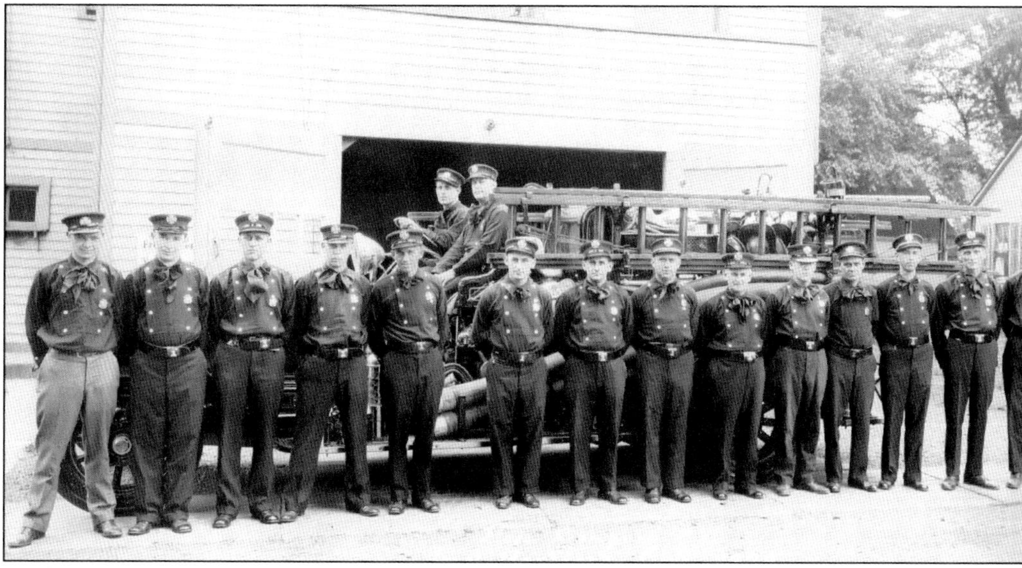

In this 1929 view are the firefighters of Engine Company No. 1 and their 1917 American LaFrance pumper. This truck was replaced in 1929 with a new Mack 1,000-gallon-per-minute pumper. (Courtesy of the East Hartford Fire Department.)

Report of Chief of Fire Department

To the Board of Commissioners of the East Hartford Fire District:

I beg leave to submit the following report of the East Hartford Fire Department for the year ending May 7, 1918.

During the year the Department responded to the following alarms:

1917		
May 5	Still Alarm	H. Wehrly, Governor St., oil stove explosion, damage small.
May 29	Box 27	Mrs. Mary Kilgariff, Woodbridge Ave., loss $400, insurance.
June 10	Box 16	Carbarn dump, no damage.
June 21	Box 35	Rear pool room, Burnside, no damage.
July 3	Box 12	Hunting Bros., Burnside Ave., loss $500, insurance.
July 9	Box 16	Carbarn dump, no damage.
July 17	Box 12	T. M. Noble, Main St., loss $50, insurance.
July 19	Box 14	Auto on fire.
July 21	Still Alarm	Carbarn dump, no damage.
Aug. 6	Box 26	Est. of A. S. Bailey, rear 40 Ranney St., loss $100, no insurance.
Aug. 25	Still Alarm	H. W. Grant, Bissell St., hen coop, damage small.
Sept. 4	Box 32	Mrs. J. Price, Tolland St., loss $1,100, insurance.
Oct. 3	Box 32	Mrs. Mary Lyons, Burnside Ave., loss $140 insurance.
Oct. 22	Box 351	Myrtle Ave., grass fire, no damage.
Oct. 29	Box 251	Sta. 22½, Burnside Ave., no damage.
Nov. 10	Still Alarm	Brush fire, no damage.
Nov. 12	Box 15	Grass fire, no damage.
Nov. 18	Box 27	Chimney fire, Ranney St., no damage.
Dec. 6	Box 24	Grass fire, Prospect St., no damage.
Dec. 27	Box 12	James A. Martin, hen house, damage small.
1918		
Jan. 11	Box 321	105 Larrabee St., no damage.
Jan. 28	Box 26	False alarm.
Jan. 30	Box 321	62 Bliss St., gas stove explosion, damage small.
Feb. 1	Box 14	Chimney fire, C. F. Hamner, Main St., no damage.
Feb. 4	Box 26	Chimney fire, Ranney St., no damage.
Feb. 18	Box 123	Trolley car fire.
Mar. 6	Box 213	Chimney fire, Huntting Bros., Burnside Ave.
Mar. 19	Box 35	F. Conwell, tobacco shed, loss $600, insurance $200.
Mar. 28	Box 26	No fire.
Mar. 29	Box 35	Grass fire.
Mar. 29	Still Alarm	Grass fire.
April 11	Box 36	Richmond Park Ave., chimney fire.
April 22	Box 15	Automobile on fire.
April 24	Box 12	Rubbish rear T. Noble store.
May 5	Box 25	Michael Ravelle, Melrose St., loss $50, insurance.

We have responded to 7 out of district alarms for assistance.

Respectfully submitted,

EDWARD BRAGG,

Chief of East Hartford Fire Department.

The 1918 annual report of the East Hartford Fire District shows all the fires it responded to during the period of July 1917 to June 1918. (Courtesy of the East Hartford Public Library.)

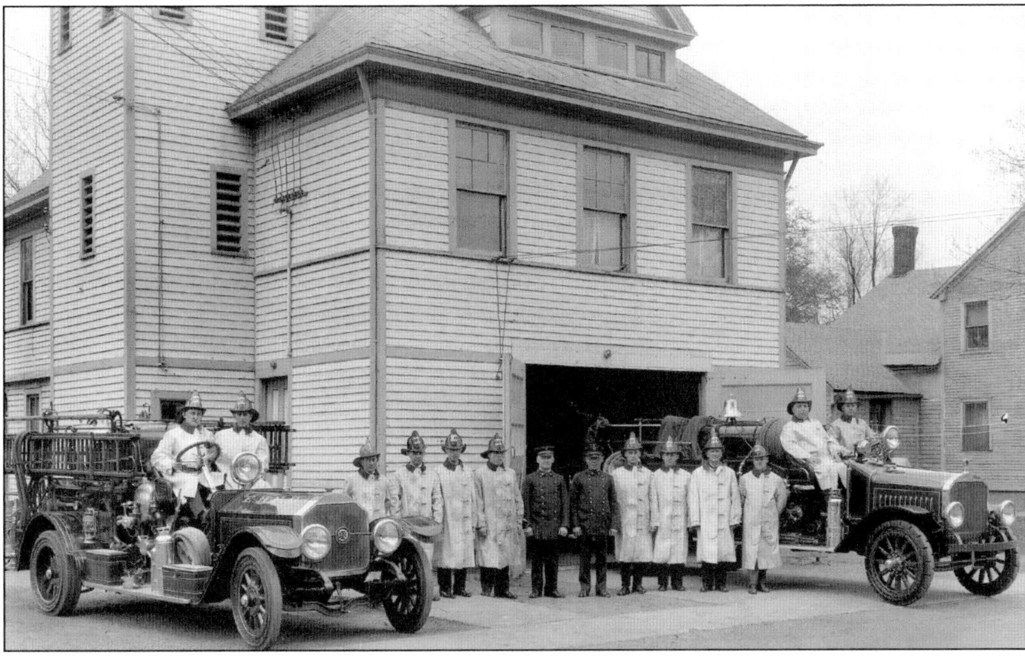

Center Hose Company station shares this 1925 scene with Engine No. 1's 1917 American LaFrance and Engine No. 3's 1923 Mack pumper, along with the department of 12 firefighters and Assistant Chief John Armstrong and Chief Frank Meunier. (Courtesy of the East Hartford Fire Department.)

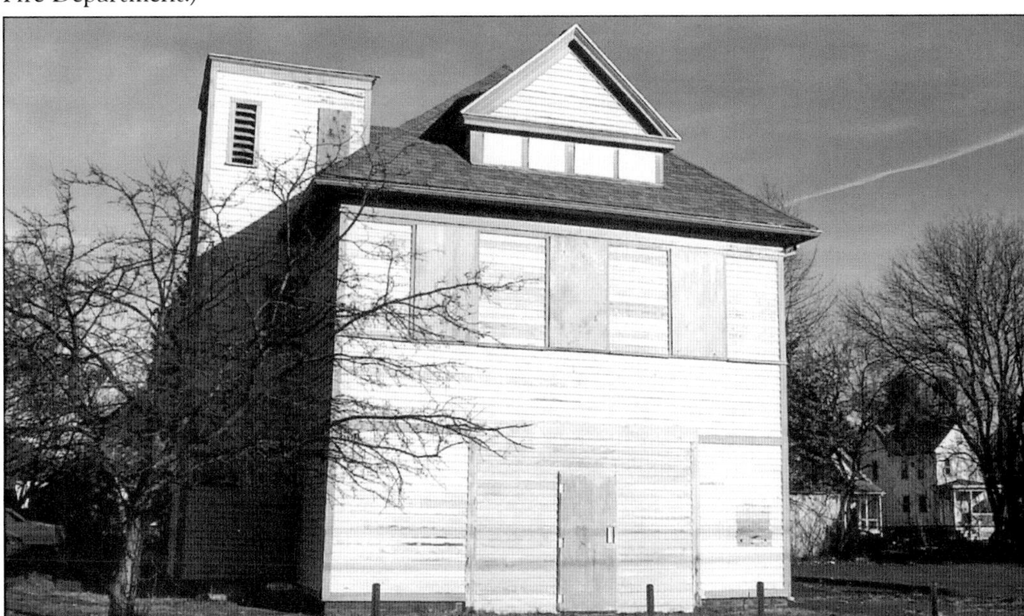

In this view is the Center Hose Company's Bissell Street station in 2005. Prior to motorization, this station had a horse-drawn hose and ladder wagon. Rather than keep the horses in the station, they were boarded across the street in a livery stable. As of 2006, a group of East Hartford firefighters hope to convert this building into a museum of firefighting history. (Courtesy of the Connecticut Fire Museum Archives.)

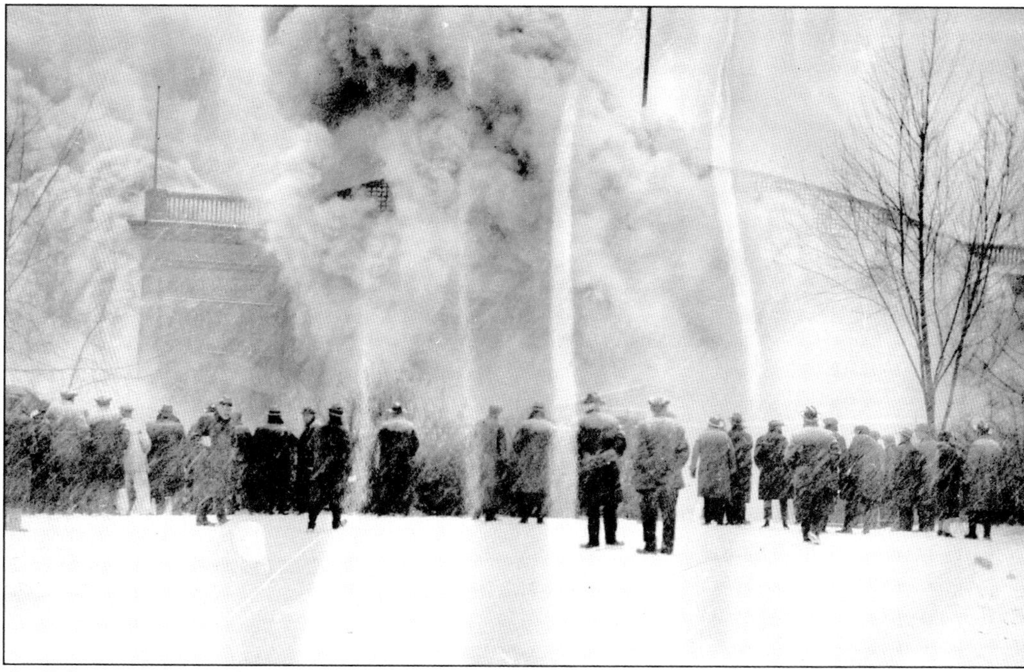

On February 1, 1926, the La Bal Tabarin dance hall, near present-day Darlin Street, burned to the ground in a spectacular fire that required mutual aid from the Hartford and South Manchester departments. (Courtesy of the East Hartford Fire Department.)

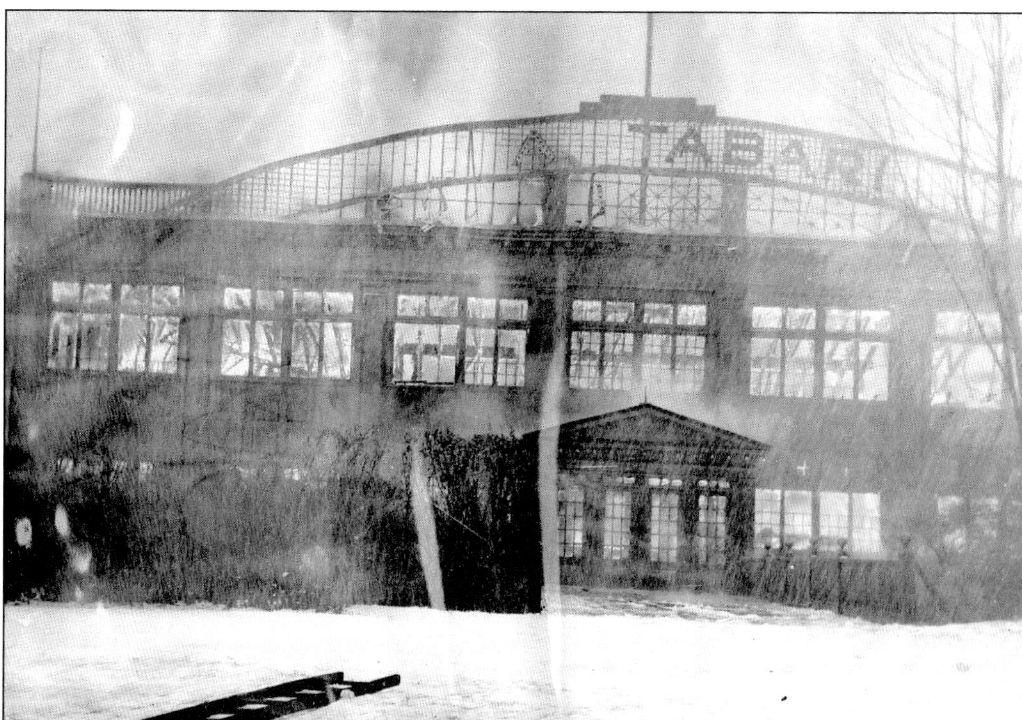

This is another view of the La Bal Tabarin fire. (Courtesy of the East Hartford Fire Department.)

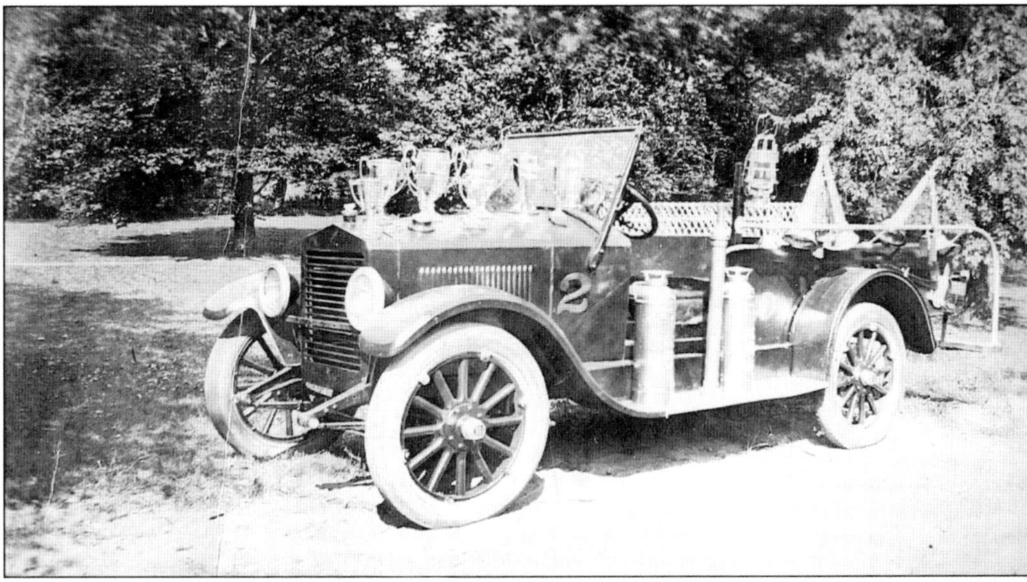

Volunteer Hose Company No. 2's REO muster wagon is seen here. From the number of trophies on the hood, it looks like the company did well at competitions. Fire musters were held by volunteer departments to test the skill of the members and to compete with other departments. (Courtesy of the East Hartford Fire Department.)

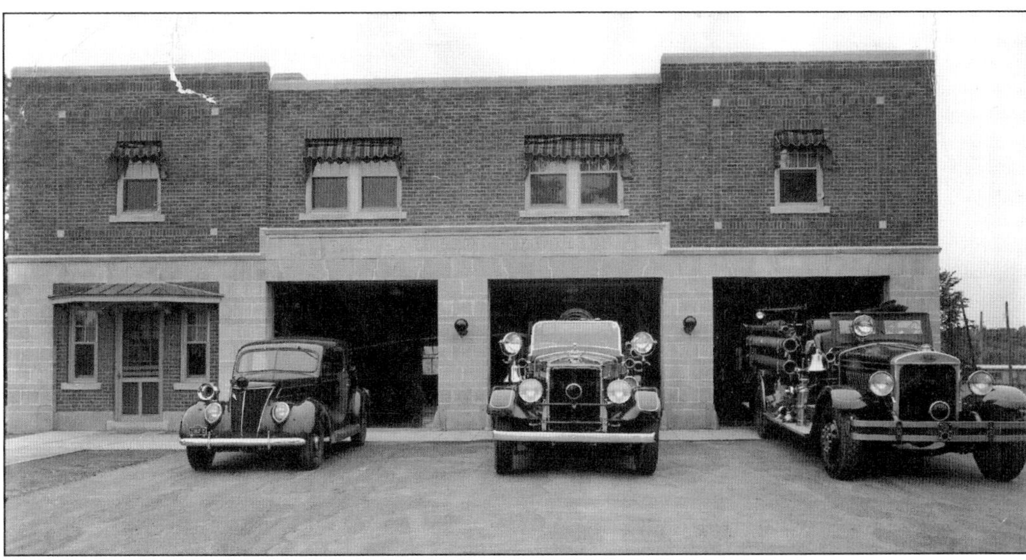

East Hartford's new station for Company No. 1 is shown in this 1938 view with Engine Company No. 2's 1934 Mack pumper and Engine 1's 1929 Mack pumper. This building was opened in 1937 to replace the original Center Hose building on Bissell Street built in 1893. This building was financed with the aid of the federal government through the Public Works Administration. This was formally the site of the Connecticut Company carbarn. (Courtesy of the East Hartford Fire Department.)

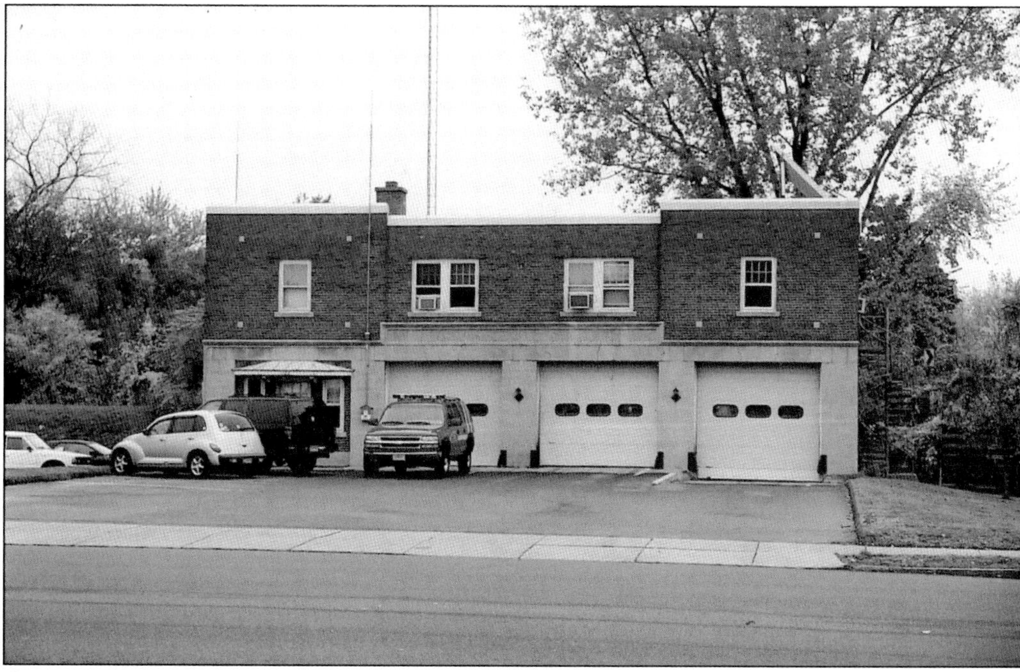

Company No. 1's quarters are seen in this 2006 view. In 2003, fire headquarters moved to the new School Street station of Company No. 3. (Courtesy of the Connecticut Fire Museum Archives.)

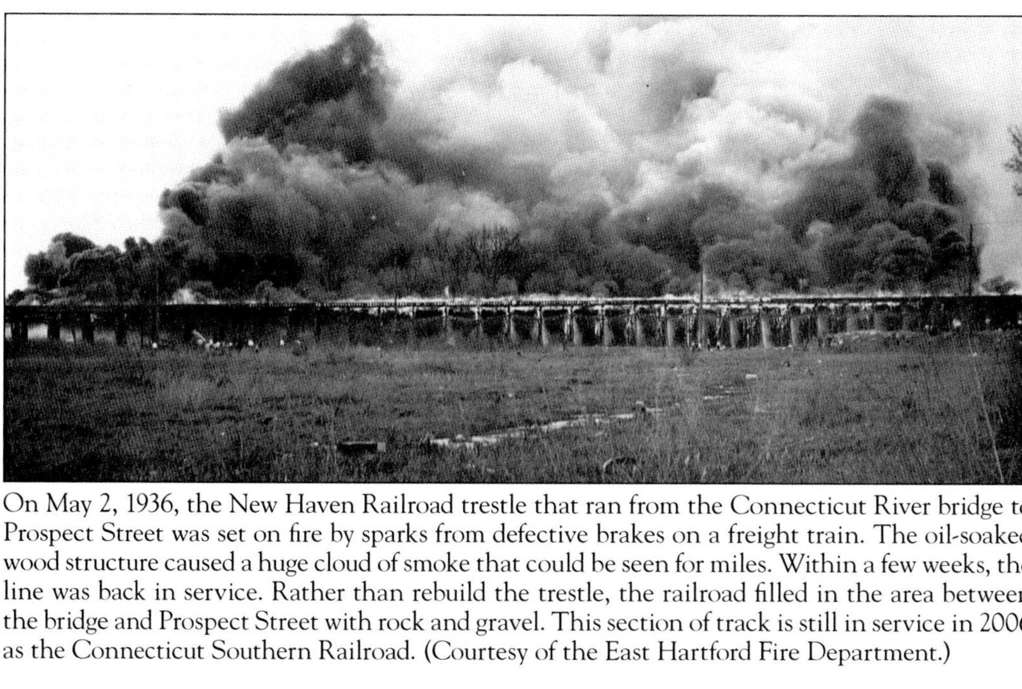

On May 2, 1936, the New Haven Railroad trestle that ran from the Connecticut River bridge to Prospect Street was set on fire by sparks from defective brakes on a freight train. The oil-soaked wood structure caused a huge cloud of smoke that could be seen for miles. Within a few weeks, the line was back in service. Rather than rebuild the trestle, the railroad filled in the area between the bridge and Prospect Street with rock and gravel. This section of track is still in service in 2006 as the Connecticut Southern Railroad. (Courtesy of the East Hartford Fire Department.)

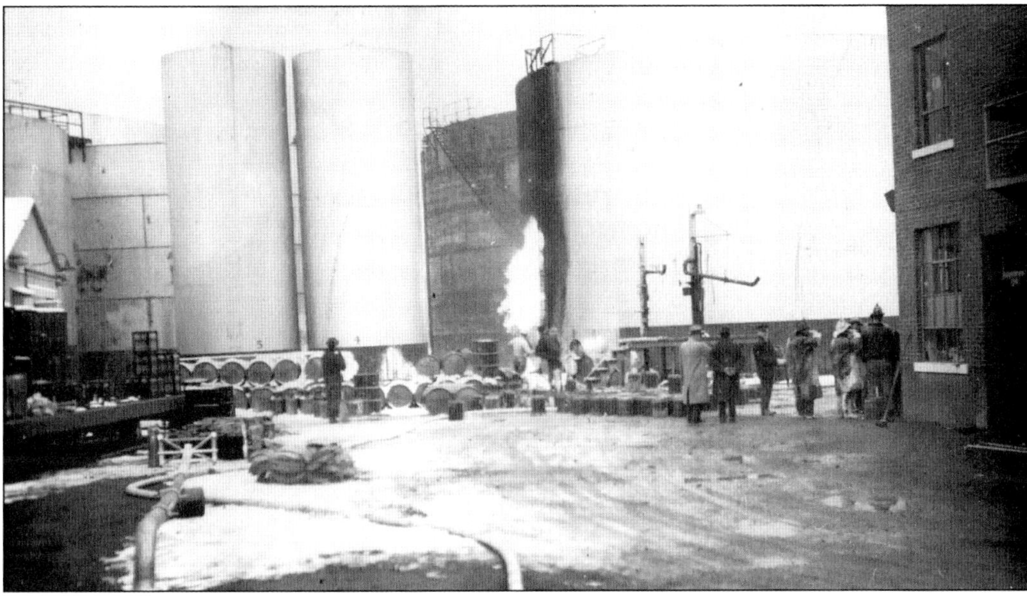

The Shell Oil Company depot on Main Street burned on November 25, 1936. Petroleum fires were particularly hard to put out and usually required large quantities of foam to bring the fire under control. Firefighting foam consists of a mass of fine bubbles that cools the fire and removes the necessary oxygen for the fire. East Hartford did not have a piece of apparatus dedicated to petroleum-type fires until 1946 when a war surplus Chevrolet foam truck was purchased. (Courtesy of the East Hartford Fire Department.)

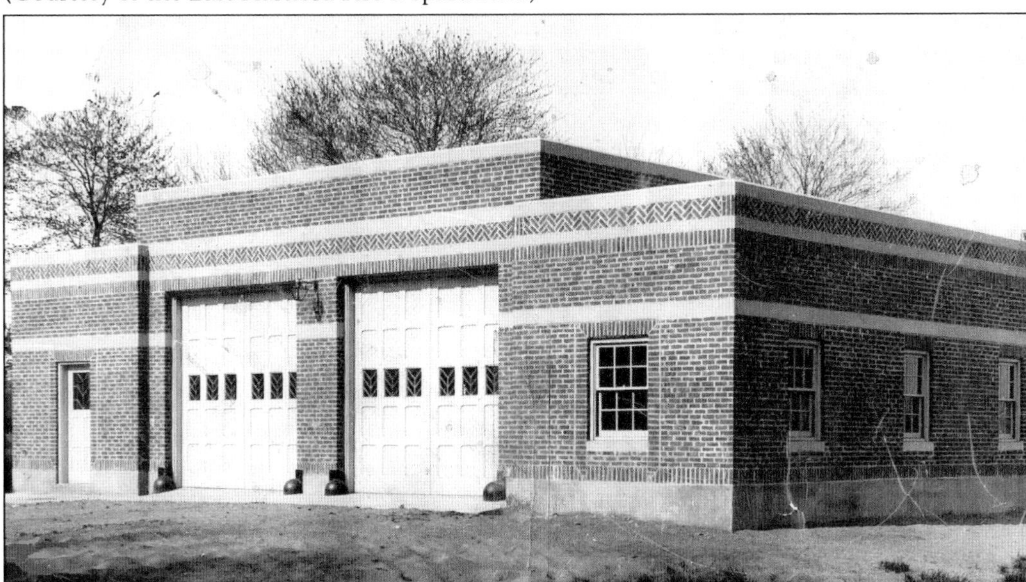

Engine Company No. 5 was organized in 1931 to provide protection to the south end of town. Originally built as a one-story building, a second floor was added in the late 1930s. Engine No. 5's first fire apparatus was a 1917 American LaFrance pumper that originally served Center Hose Company. This company was organized as a paid station and only had volunteers for a short time in the 1930s. Like the other companies, it did have paid "callmen" that responded only for fires and did not normally stay in the station. (Courtesy of the East Hartford Fire Department.)

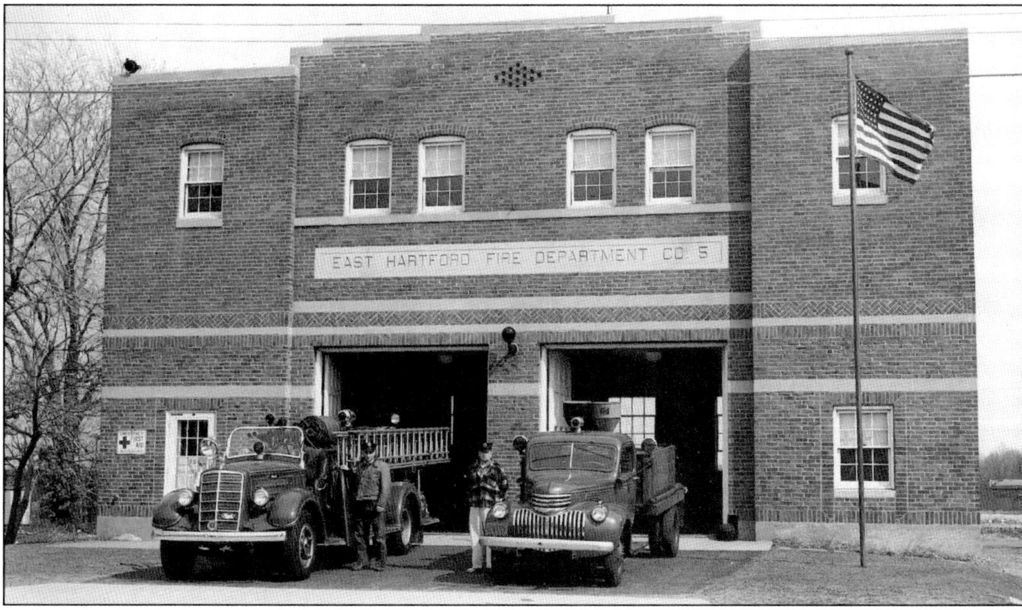

Seen in this April 1954 view is Engine No. 5's 1940 Mack pumper and the 1941 Chevrolet foam truck. The department acquired the foam truck surplus from the U.S. Army Air Corps in 1946. The foam unit was designated as Engine No. 4 and was manned only when it was needed at a fire scene. Firefighting foam in this era was in a powder form. Today foam is almost always in a liquid form, like dish detergent but more highly concentrated. (Courtesy of the East Hartford Fire Department.)

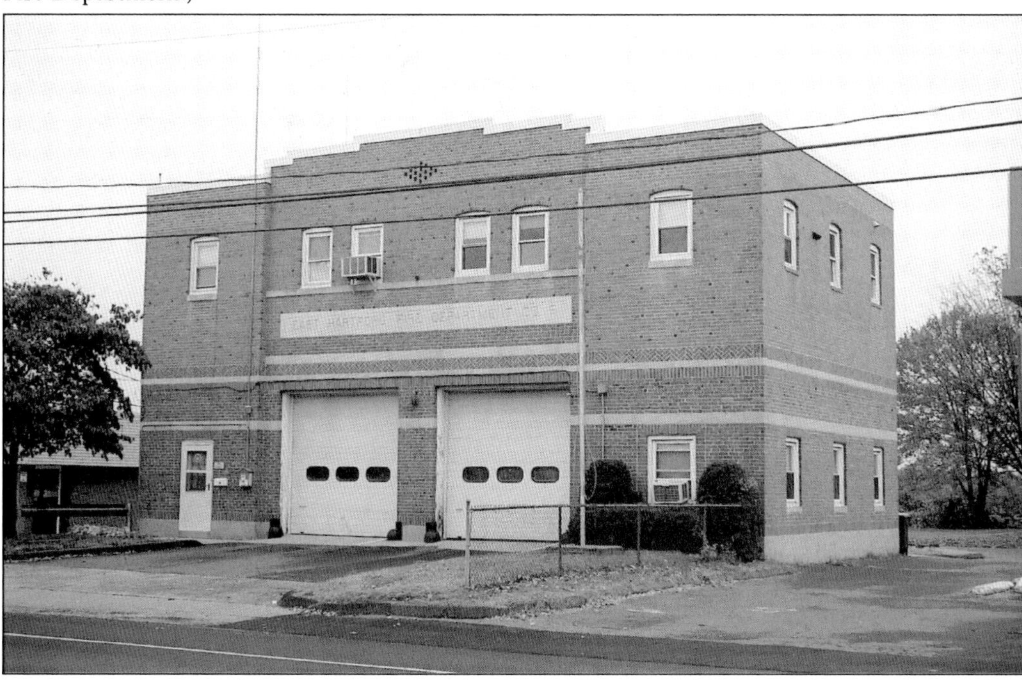

Station No. 5 is seen here in 2004. In the late 1930s, a second floor was added to the building, and the swinging apparatus doors have been replaced with more modern roll-up doors. (Courtesy of the Connecticut Fire Museum Archives.)

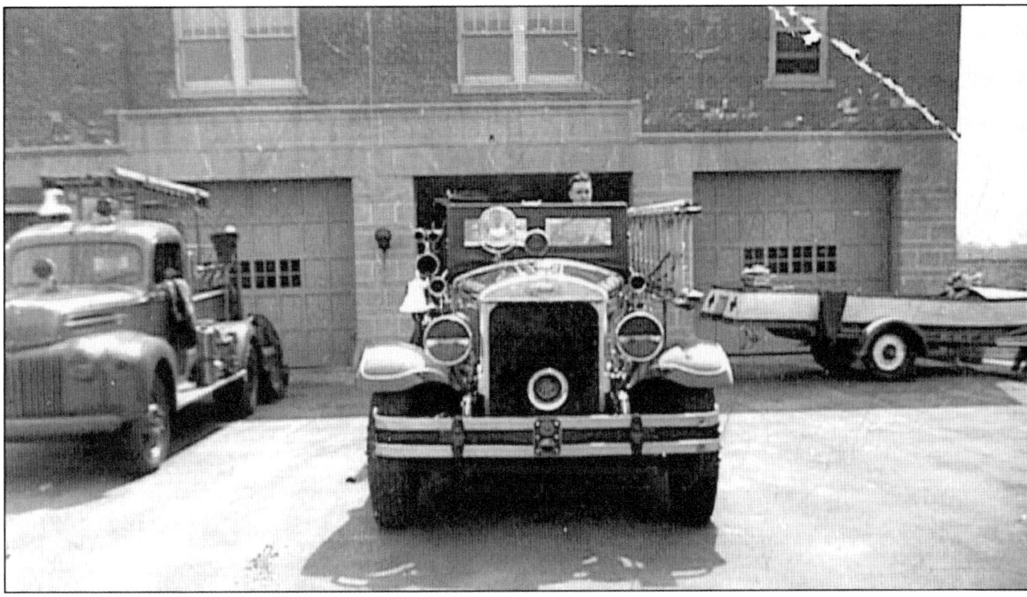

Engine Company No. 1's 1929 Mack pumper is shown in this wartime view with the "black out" headlights required from 1942 to 1945. On the left is one of the auxiliary fire service units that served in East Hartford during the war that were manned by volunteers. East Hartford had three units that were placed around town in various locations; the auxiliary volunteers were supervised by the regular paid firefighters. (Courtesy of the East Hartford Fire Department.)

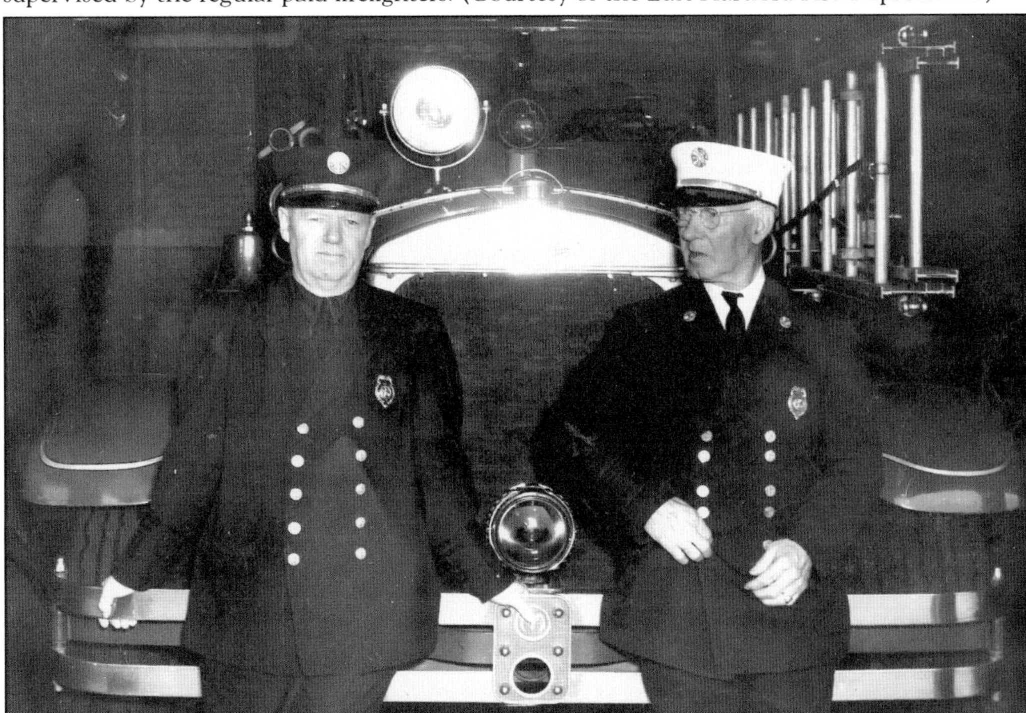

Chief John Armstrong and fire marshal Sam Clayton are seen in this 1948 view with Engine Company No. 1's Mack pumper. (Courtesy of the East Hartford Fire Department.)

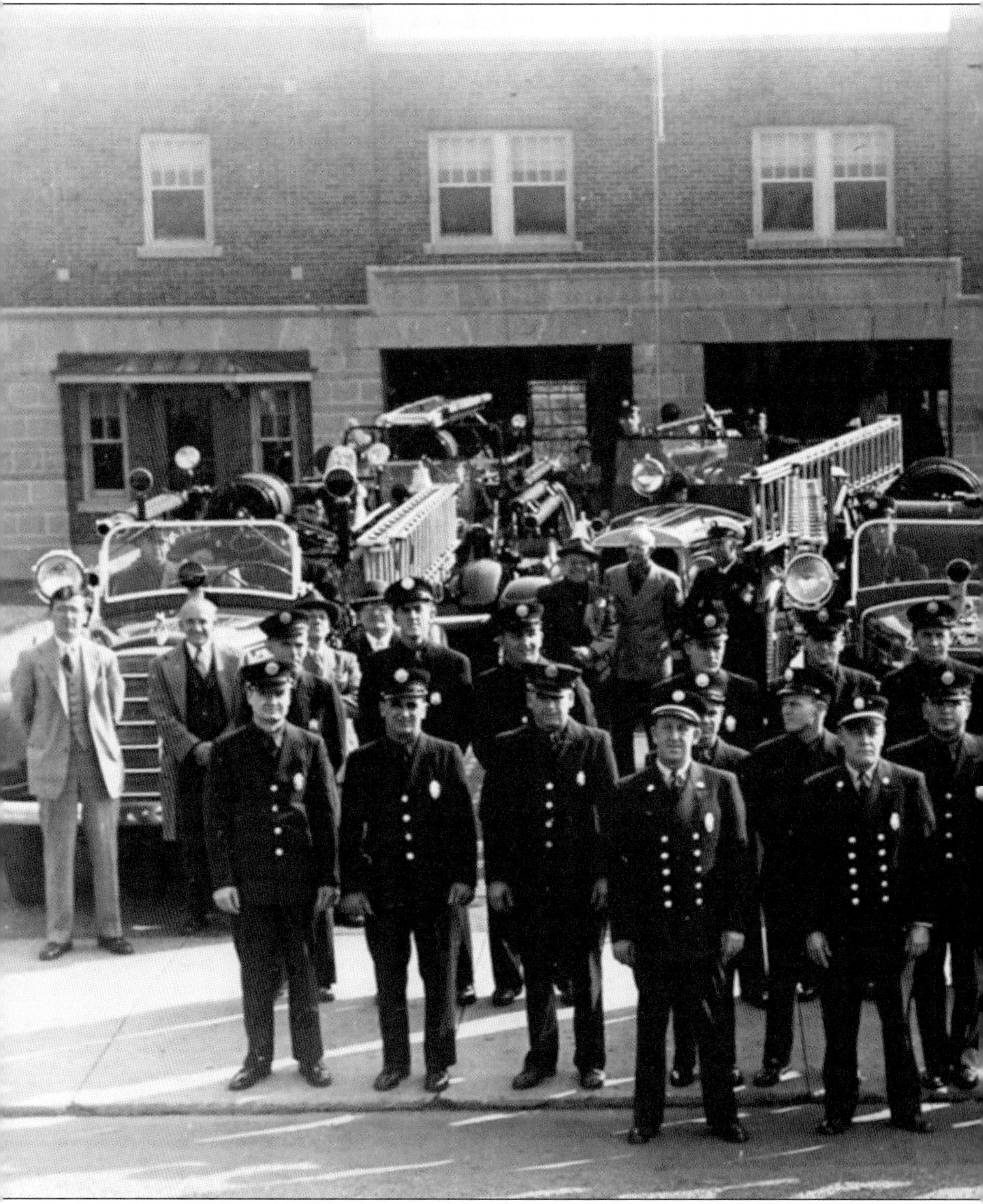

The 47 firefighters with the apparatus of the East Hartford Fire Department are shown in this 1950 view at headquarters on Main Street. In 1940, the department hired 12 new firefighters, and by 1941, the department had 23 paid men on the roster. By 1949, the department had

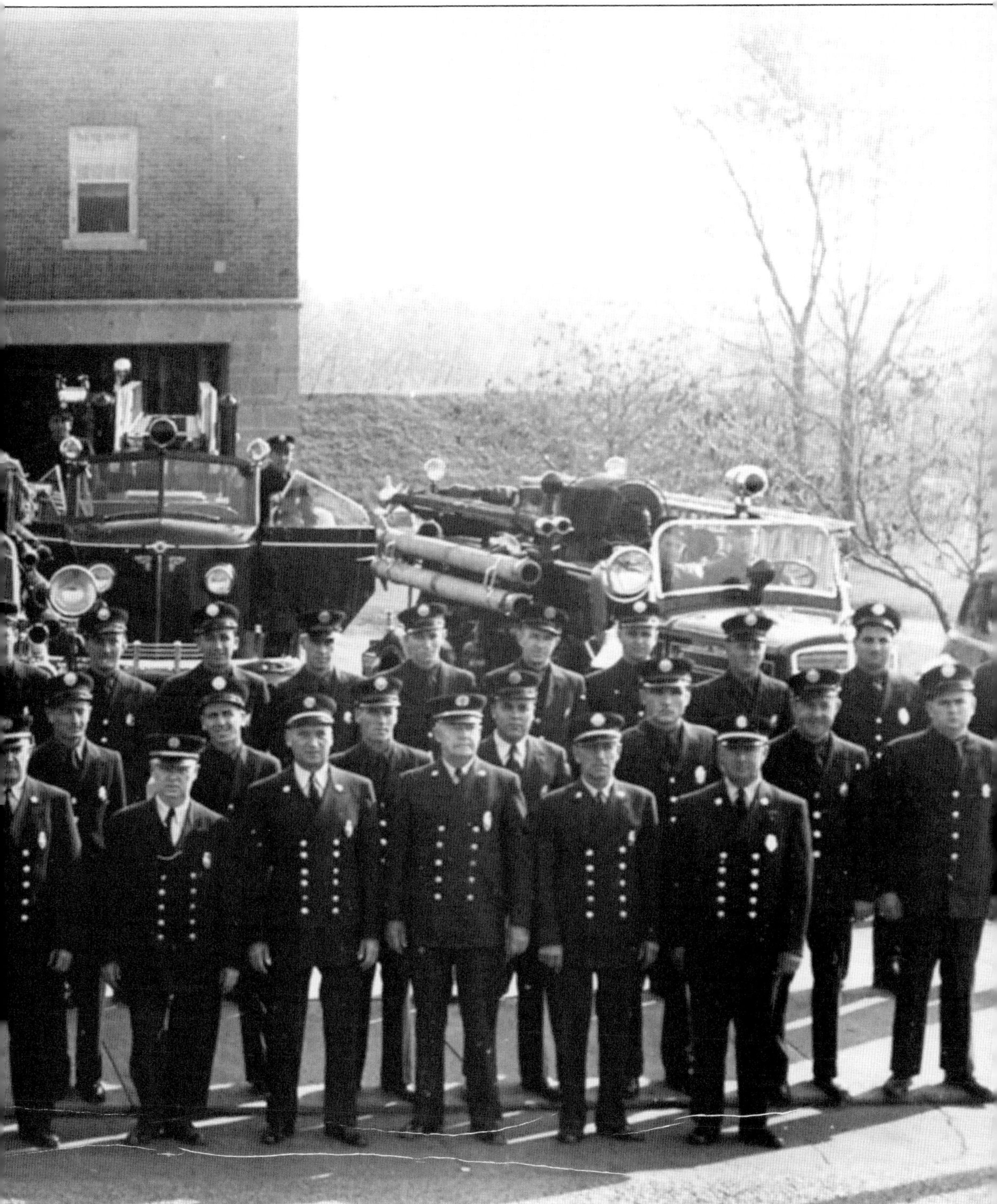

44 paid firefighters supported by 52 callmen. All the pumpers in the department were built by Mack Trucks, the 1948 American LaFrance 75-foot aerial ladder truck being the only exception. (Courtesy of the East Hartford Fire Department.)

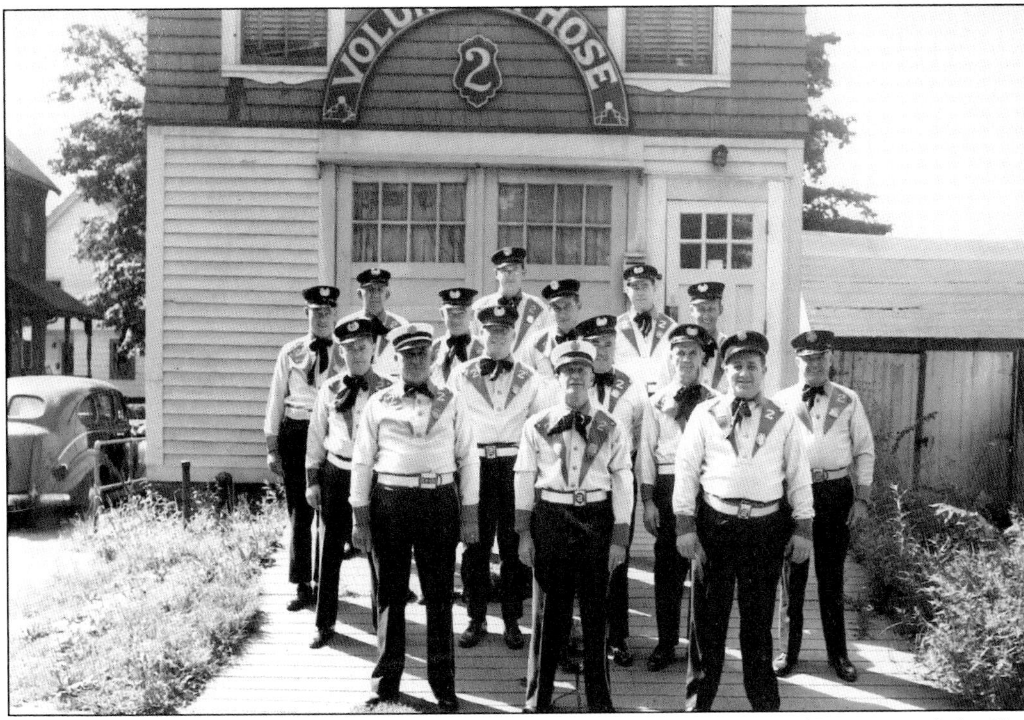

Volunteer Hose Company No. 2 members are shown in this 1949 view. (Courtesy of the East Hartford Fire Department.)

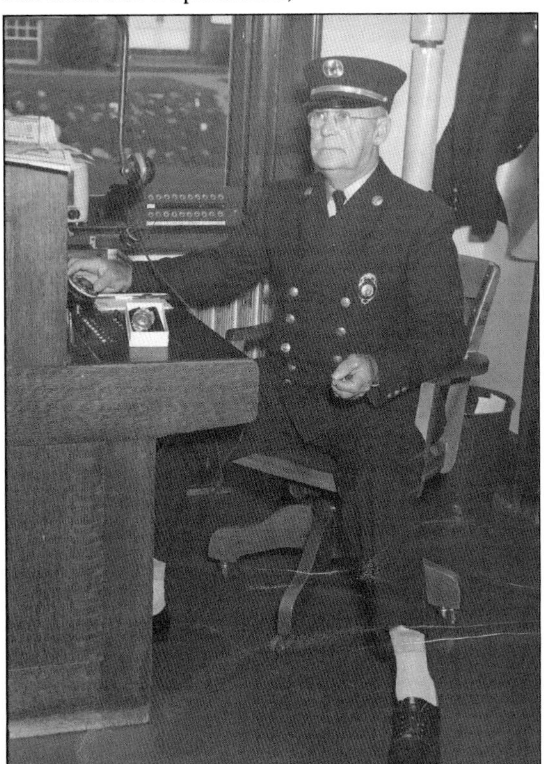

Capt. Thomas Moore is shown at the telephone switchboard at Company No. 1 in the late 1940s. Telephones were first installed in East Hartford in 1883, but most were in businesses, and many homes did not have telephones well into the 1920s. Many alarms came into the office on the fire alarm telegraph system from one of the street boxes located on the major streets. Until two-way radios became available after World War II, the telephone was the primary form of communication. Not until the late 1940s did fire apparatus have radios to keep in contact with the alarm center while out of the station, something people might take for granted today. In the early years of radio communication, the fire and police departments shared one VHF frequency. In later years, the fire department received its own VHF channel. In the 1970s, East Hartford was the first department in Connecticut to switch over to a UHF channel for radio communication. (Courtesy of the East Hartford Fire Department.)

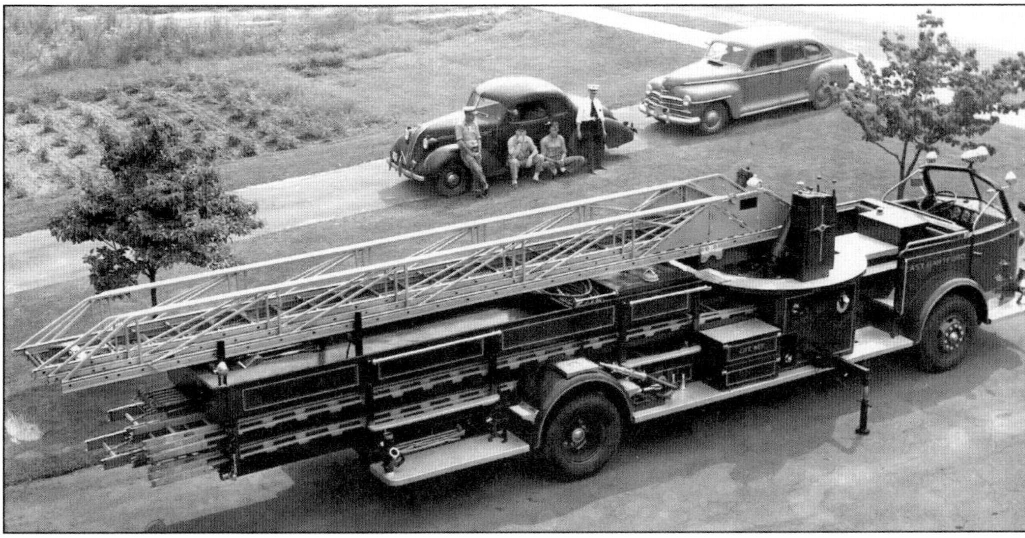

East Hartford's new 75-foot American LaFrance aerial ladder is shown in this July 1948 photograph, just after delivery. The truck was ordered in 1946 at a cost of $24,386. In 1926, Chief Frank Meunier had recommended that the fire district purchase a ladder truck "as with the present equipment rescue work can be done only up to a 3 story building." This was the only ladder truck in town until a 100-foot aerial was purchased for Ladder No. 1 in 1968. This truck then became Ladder No. 2 when a second truck company was organized at Station No. 3. In 1984, this truck was sold to the Warehouse Point Fire District in East Windsor, where it served until it was retired in 2003. It is now on display at the Connecticut Fire Museum in East Windsor. (Courtesy of the Hartford Courant.)

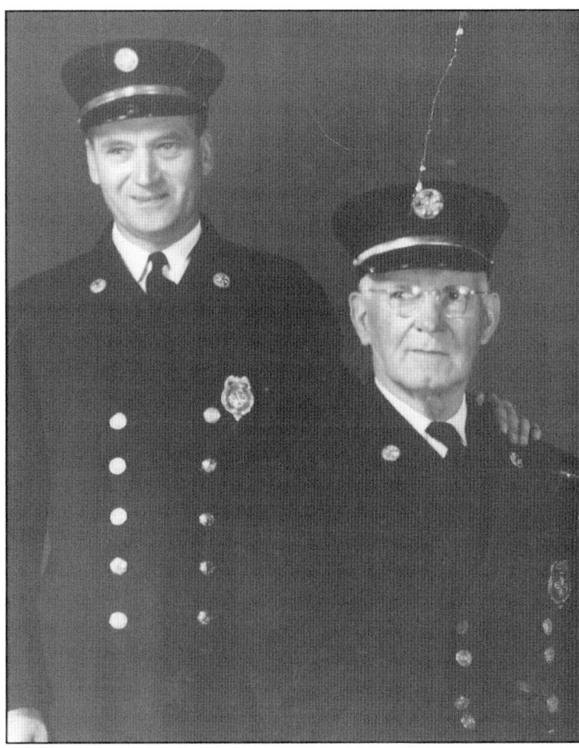

Assistant Chief Frank Dagon (left) succeeded Chief John Armstrong in 1949 and served as chief for 14 years. (Courtesy of the East Hartford Fire Department.)

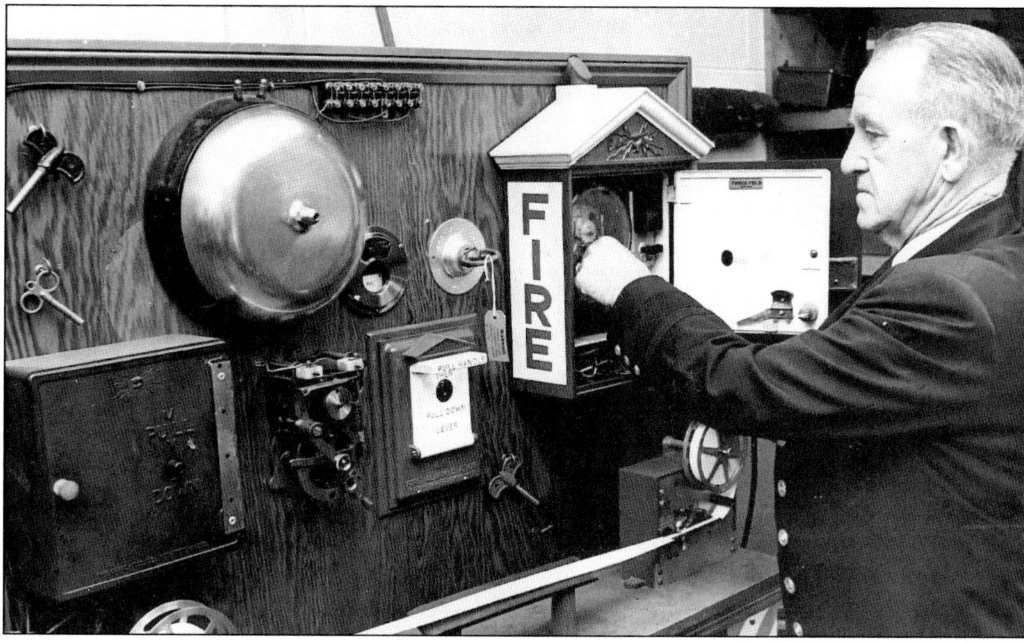

Fire alarm superintendent Henry Dawson retired from the department in 1973 after 45 years on the department. The display he is looking at was used to educate schoolchildren on the workings of the fire alarm telegraph system. East Hartford had 85 fireboxes located along all major streets, inside schools, and in large mercantile facilities. In the early 1990s, the boxes were removed from the streets, and radio boxes that did not require a substantial investment in overhead wires took their place. (Courtesy of the East Hartford Fire Department.)

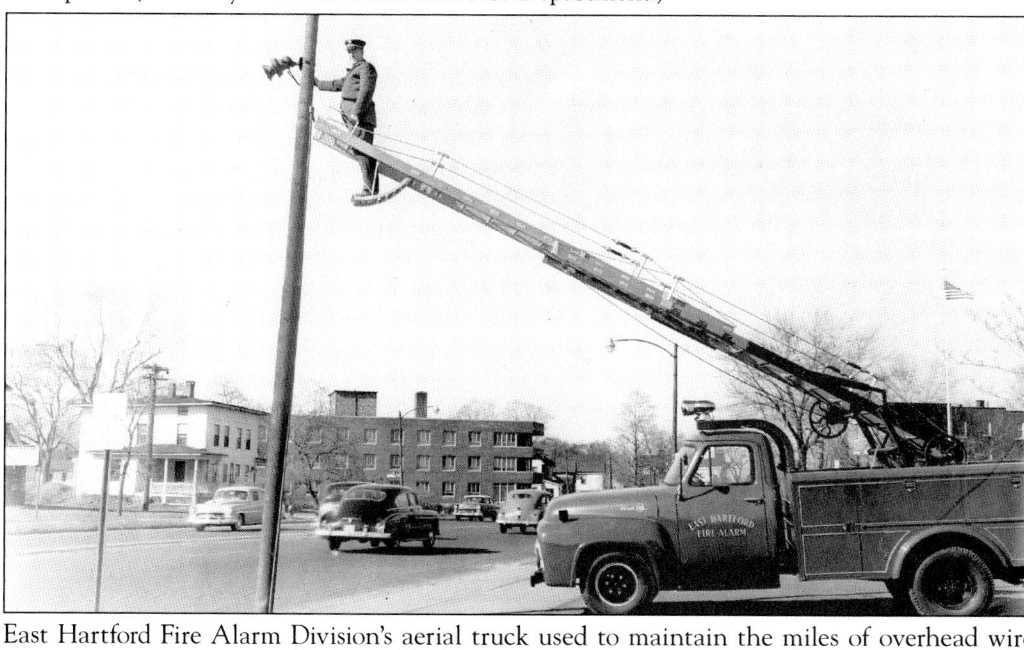

East Hartford Fire Alarm Division's aerial truck used to maintain the miles of overhead wire around town. The metal pole that holds the siren used to alert motorists when the fire apparatus is leaving the station was formally used to hold the overhead wire for the streetcar line that ran past Station No. 1 to Glastonbury. (Courtesy of the East Hartford Fire Department.)

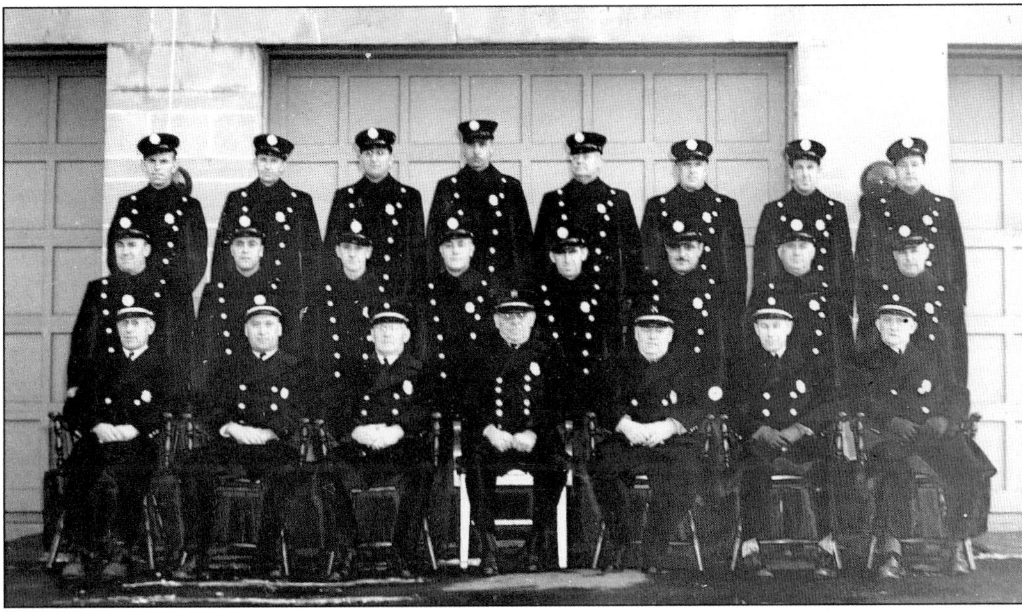

East Hartford Fire Department firefighters and officers are seen here on January 27, 1943. From left to right are (first row) Jacob Suter, Frank Dagon, John J. Armstrong, Chief Frank Meunier, Sam Clayton, Theodore Shook, and Chris McLaughlin; (second row) Mike Fitzgerald, George Chaves, Louis Hooper, John Kelley, John Moynihan, Tony Cipolla, Adolph Rosenthal, and Dennis McCarthy; (third row) James Dooley, John F. Armstrong, Edward Heimer, Harry Hinkleman, Thomas Moore, George Beauchamp, Henry Dawson, and Louis Glode. (Courtesy of the East Hartford Fire Department.)

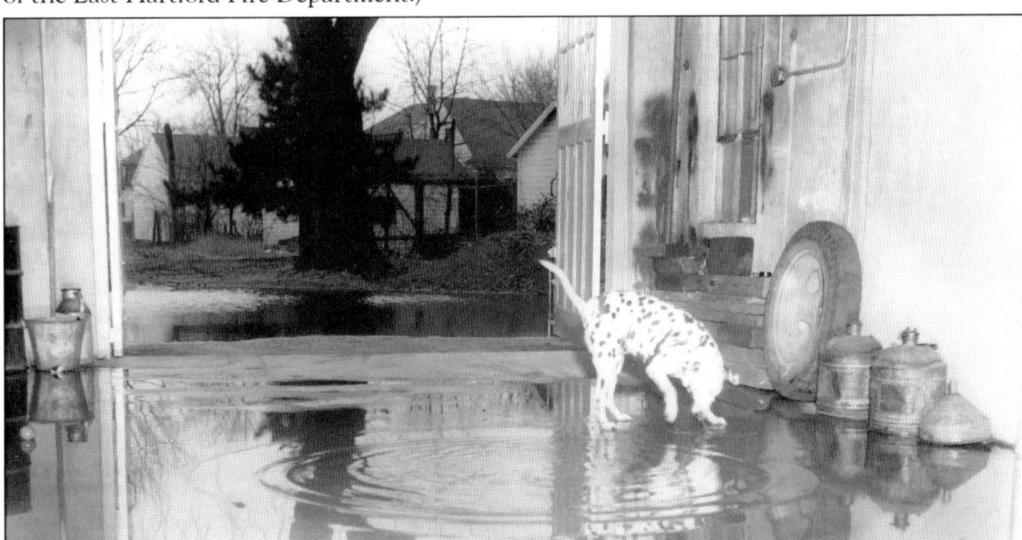

Company No. 1's dalmatian Daisy Mae is shown in this late-1940s view. Dalmatians were popular dogs around fire stations, going back to the horse-drawn days. Named after a province in Austria, these dogs have a mutual admiration with horses and were always seen in stables and firehouses. After the advent of motorized fire equipment, they are now kept as mascots and to provide security for the station when the company is out on a call. (Courtesy of the East Hartford Fire Department.)

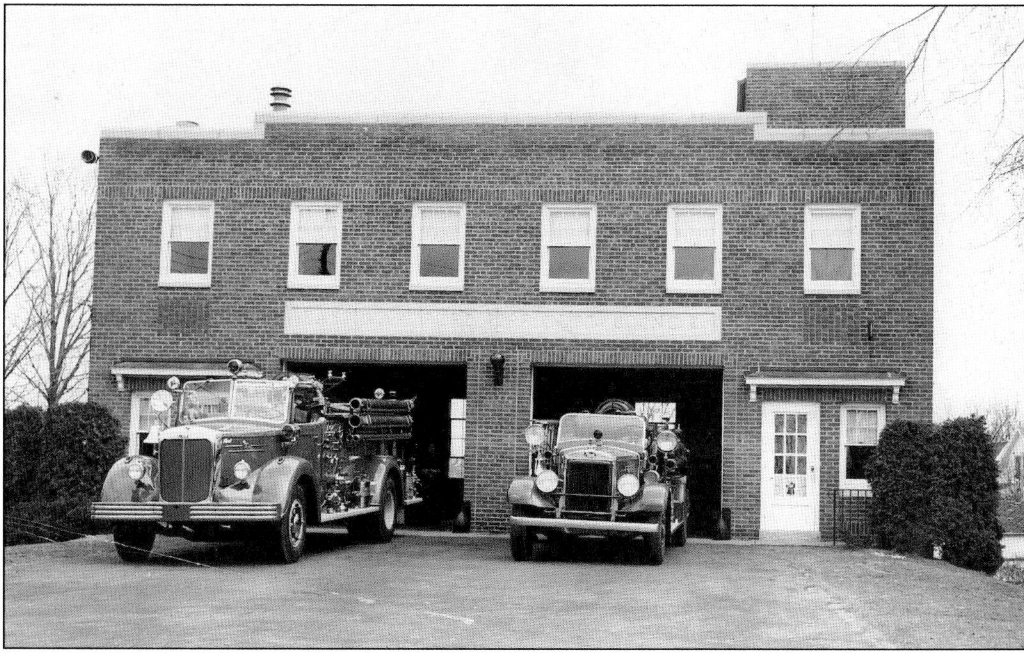

Seen here is a 1954 view of Engine Company No. 2's station at 1692 Main Street. This station was built in 1940 at a cost of $18,980. The original Hose Company No. 2 building was converted into a club for the volunteer firefighters in town. Engine No. 1's 1953 Mack type 95 pumper is alongside Engine No. 2's 1934 Mack pumper. Probably no truck builder is better known to the public than Mack, for the expression "built like a Mack truck" has become a common analogy. Mack built custom fire apparatus from 1911 to 1980. (Courtesy of the East Hartford Fire Department.)

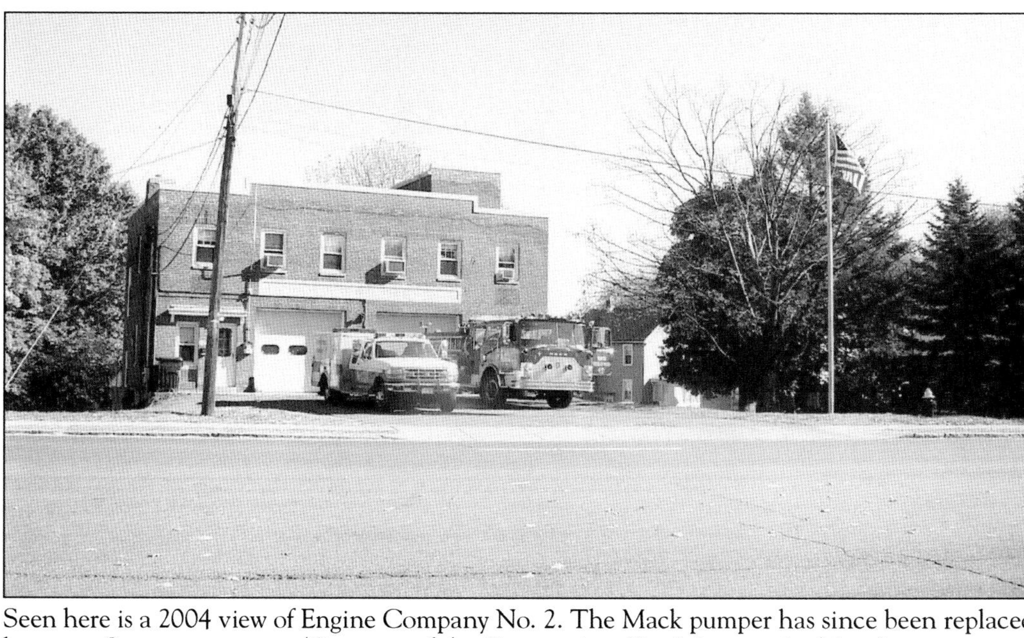

Seen here is a 2004 view of Engine Company No. 2. The Mack pumper has since been replaced by a new Seagrave pumper. (Courtesy of the Connecticut Fire Museum Archives.)

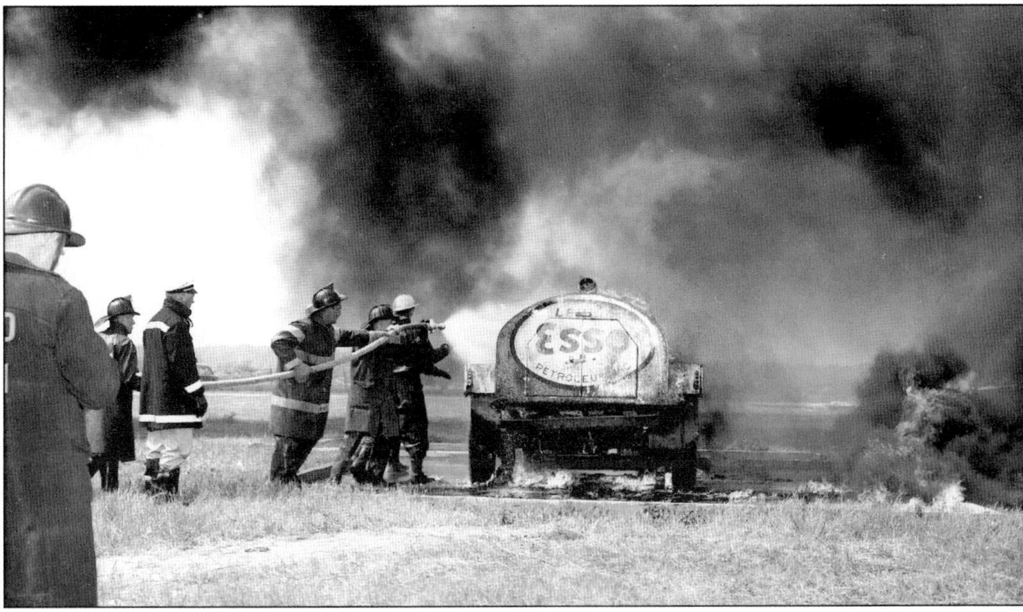

East Hartford firefighters often trained at the Pratt and Whitney Aircraft facility, where the company fire department had an area set up to simulate petroleum fires, especially aviation fuel fires. East Hartford Fire Department and the Pratt and Whitney Aircraft Fire Department have had a long history of cooperation, training, and resources that are necessary for "Hazmat" incidents in the Greater Hartford area. (Courtesy of the East Hartford Fire Department.)

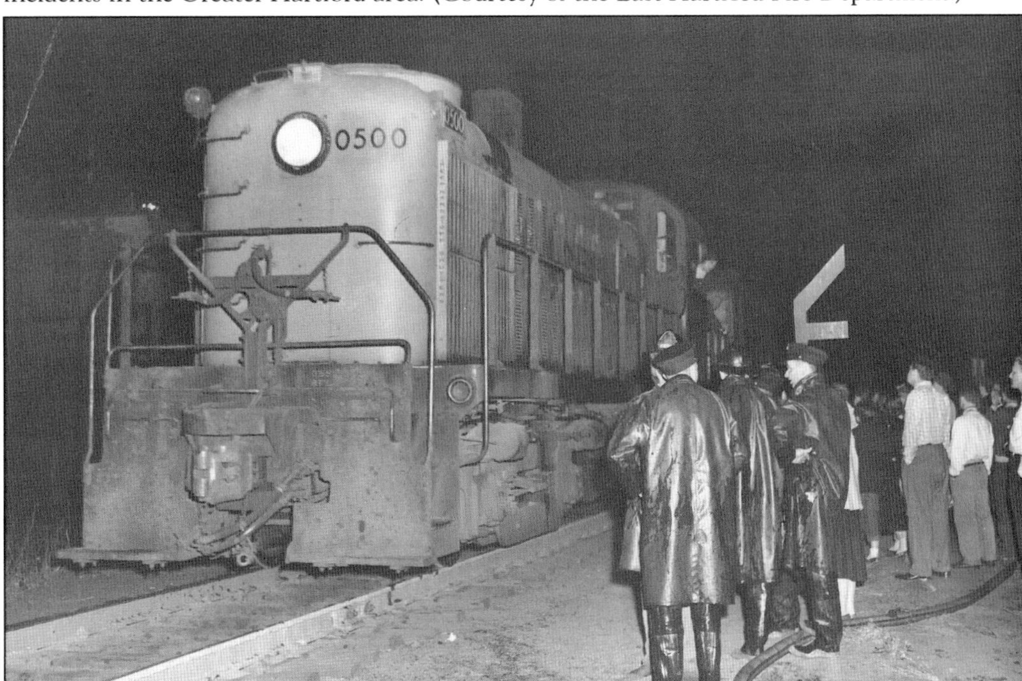

A multiple alarm on May 4, 1955, on School Street along the New Haven Railroad main line is interrupted while a Boston to Hartford passenger train passes the scene. (Courtesy of the East Hartford Fire Department.)

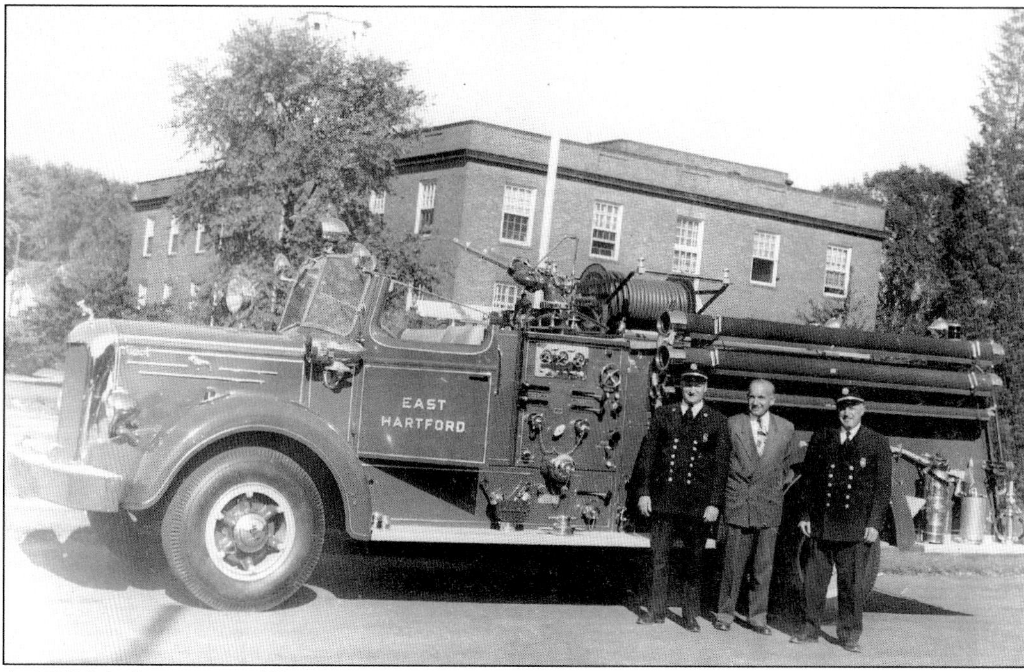

A 1953 Mack pumper of Engine Company No. 1 is shown at headquarters in the early 1950s. A department member now owns this truck. (Courtesy of the East Hartford Fire Department.)

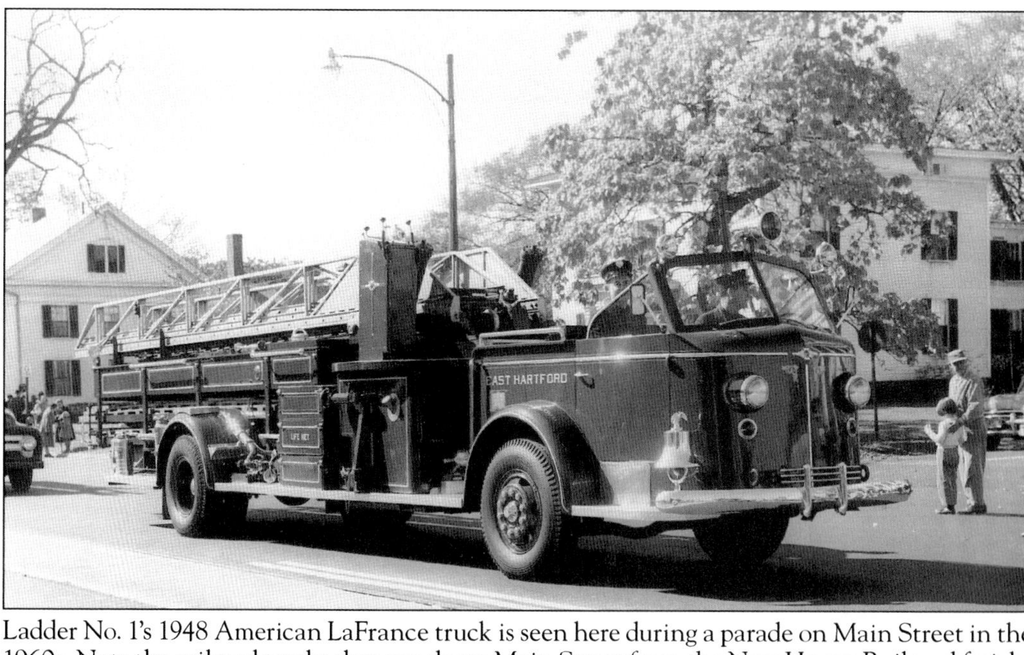

Ladder No. 1's 1948 American LaFrance truck is seen here during a parade on Main Street in the 1960s. Note the railroad tracks that ran down Main Street from the New Haven Railroad freight yard to Pratt and Whitney Aircraft about two miles away. Beyond Station No. 1, the tracks that were on a private right-of-way in the middle of Main Street made a nice private roadway for East Hartford's fire apparatus that regularly used it when responding down Main Street. (Courtesy of the East Hartford Fire Department.)

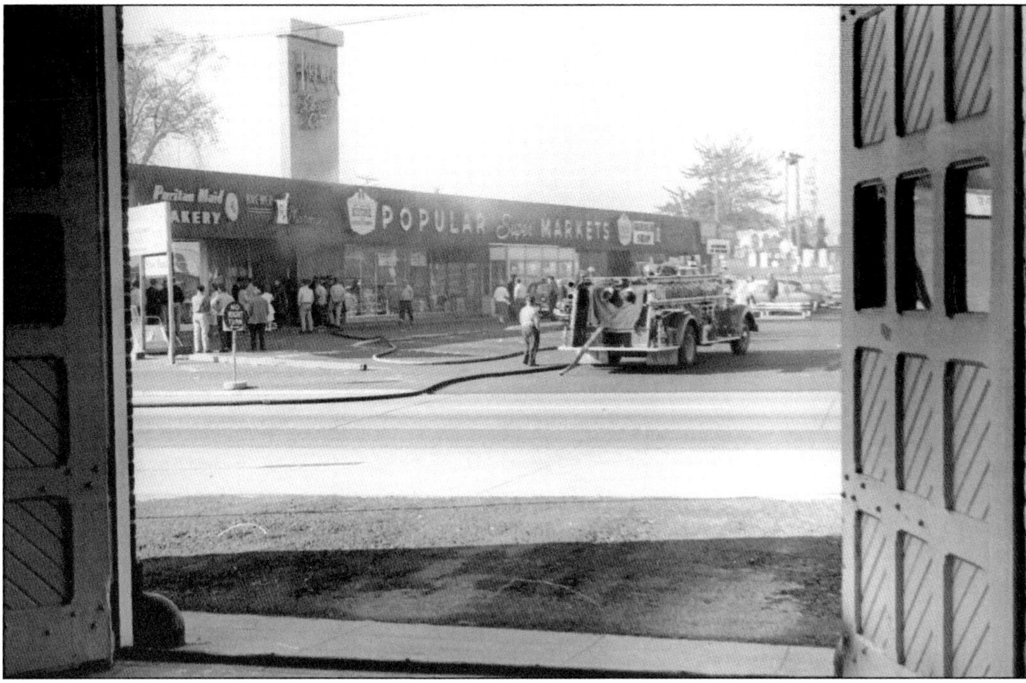

An interesting view is seen from inside Station No. 5 on Main Street. The shopping center shown in this view burned in a spectacular multiple-alarm fire in the early 1970s. (Courtesy of the East Hartford Fire Department.)

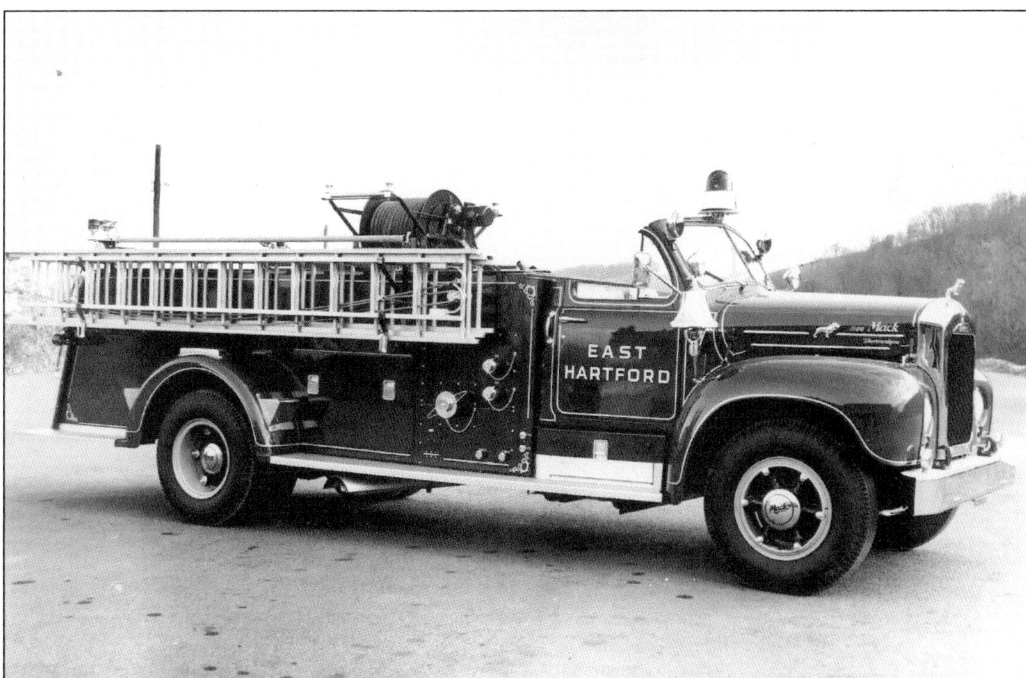

This Mack Trucks, Inc., builder's photograph of Engine No. 5's new pumper was taken at the Mack plant in Allentown, Pennsylvania. (Courtesy of Mack Trucks, Inc.)

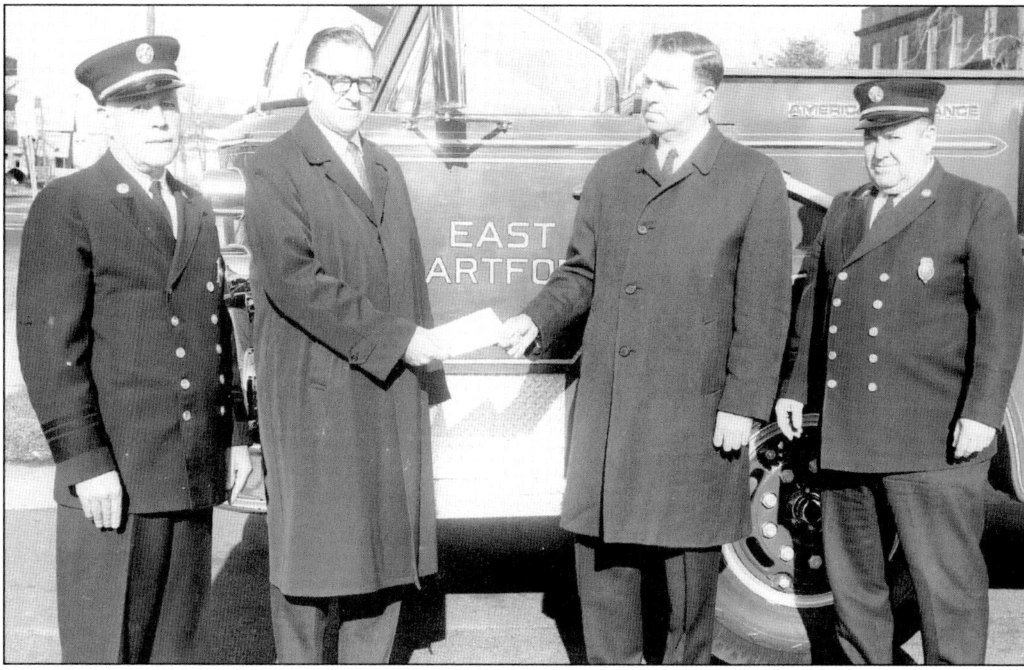

On November 17, 1968, East Hartford accepted delivery of a new 1968 American LaFrance 100-foot tillered aerial ladder truck. In this photograph are, from left to right, Chief John Kelley, Mayor Edwards Atwood, an unidentified American LaFrance representative, and Assistant Chief Mike Fitzgerald. (Courtesy of the East Hartford Fire Department.)

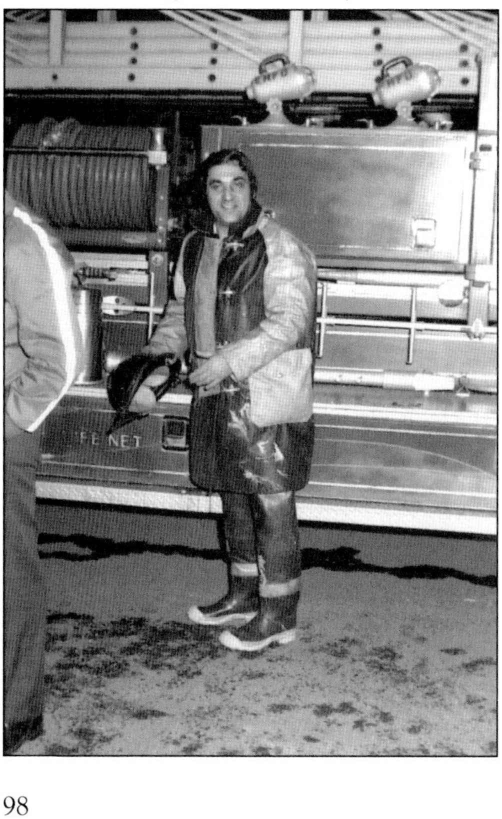

In the 1960s, firefighters still wore rubber boots and coats, as shown in this 1968 photograph of Ladder No. 1's driver, Santo Alleano, next to the 1968 American LaFrance 100-foot tiller truck. Alleano served as the International Association of Firefighters local president for many years. The firefighters' labor unions fought hard over the years to bring about the work schedules and staffing today for paid firefighters in the United States. (Courtesy of the East Hartford Fire Department.)

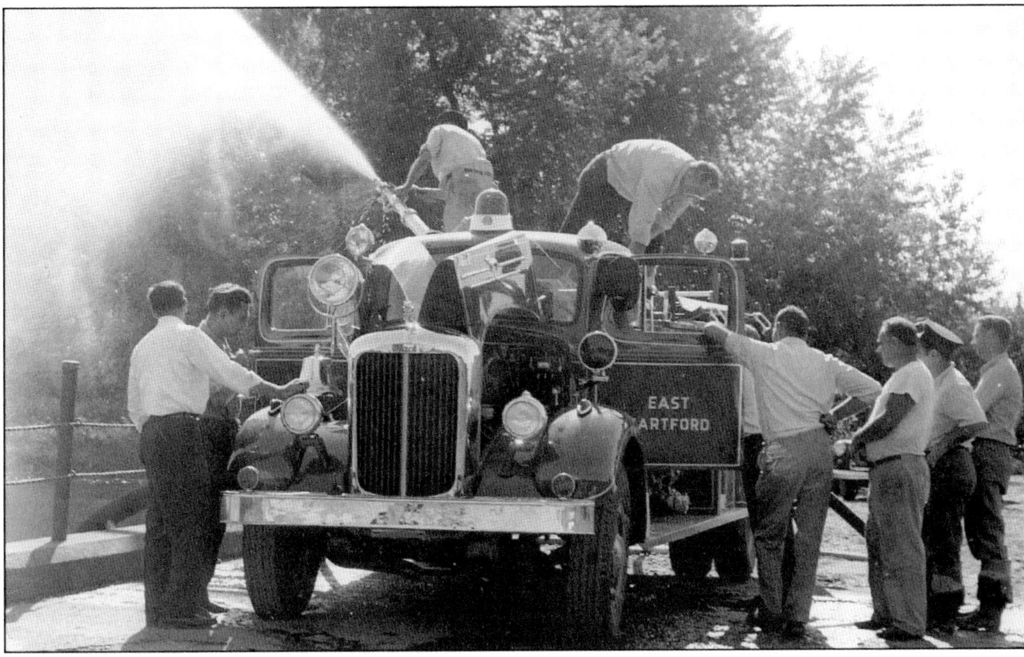

East Hartford firefighters are testing the new 1955 Mack squad truck at Keney Cove in this August 29, 1955, view. The squad truck was a familiar vehicle to residents all over town, as it answered all alarms to transport the necessary manpower to the fire scene and bring vital equipment for practically all emergencies. (Courtesy of the East Hartford Fire Department.)

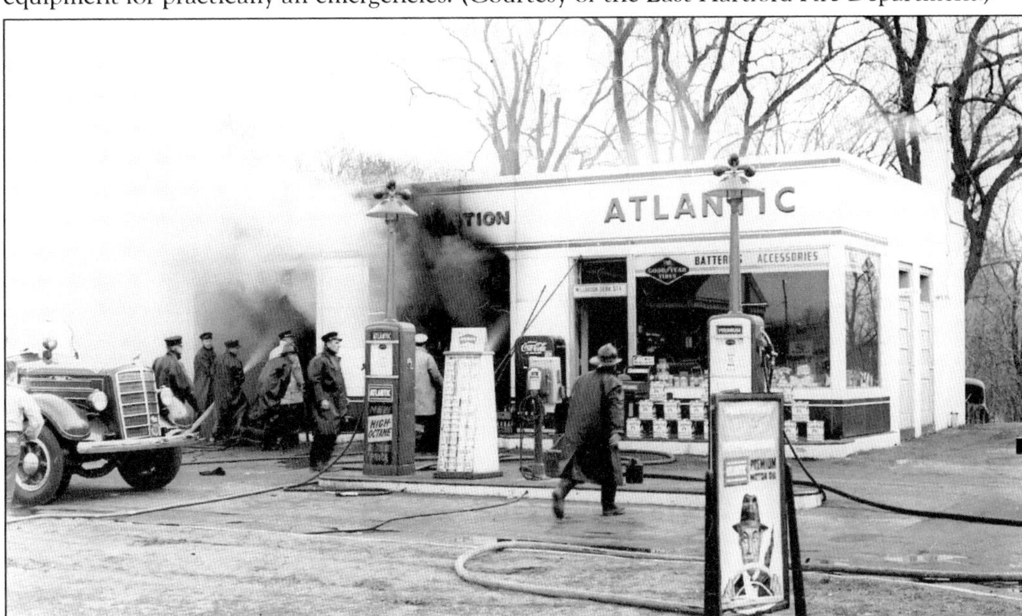

Firefighting in the 1950s was a little more informal than today, as shown in this April 22, 1956, scene. Engine No. 5's 1940 Mack has responded to the fire along with Engine No. 1's 1953 Mack pumper. The 1940 pumper, called "Maxine" by the firemen, is still owned by the department and is now in the department museum at Company No. 6 on Forbes Street. (Courtesy of the East Hartford Fire Department.)

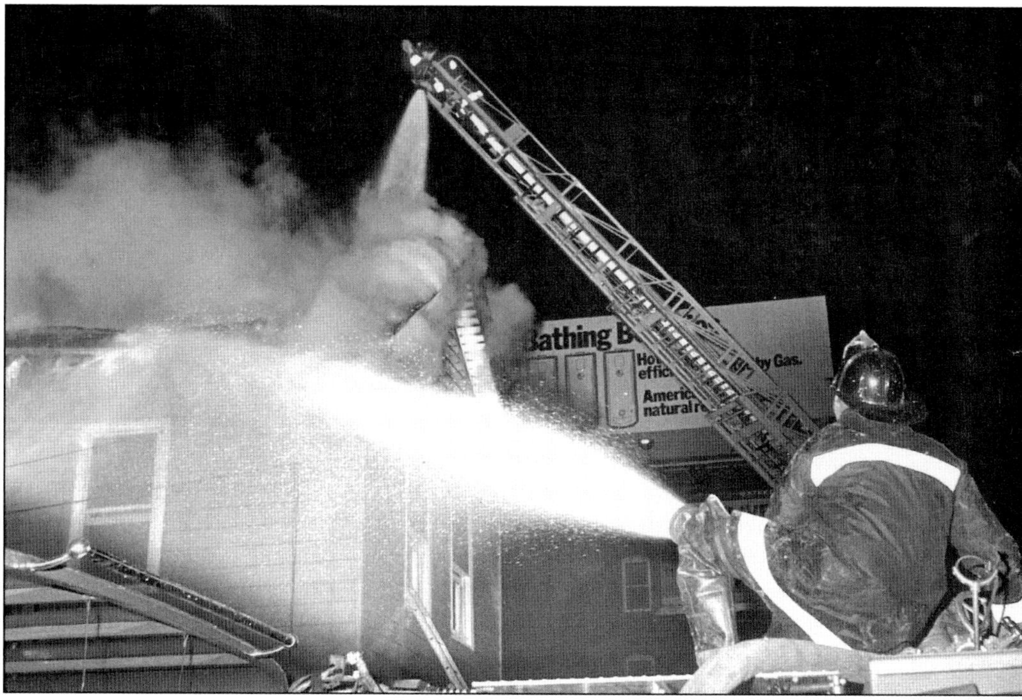

Ladder No. 1 and Engine Company No. 1 are putting water on this building fire on Burnside Avenue in the early 1970s. (Courtesy of the East Hartford Fire Department.)

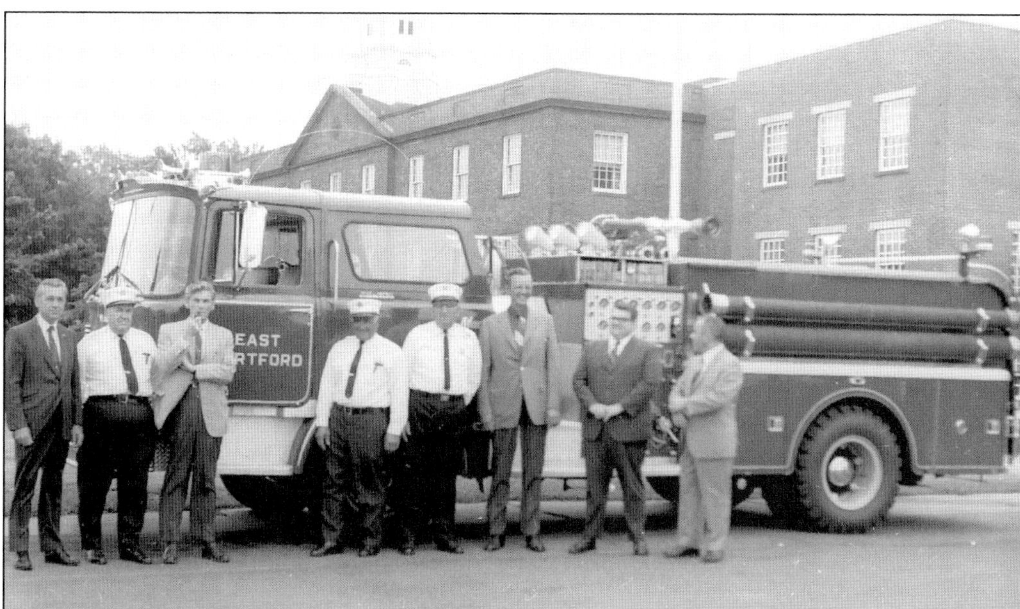

Engine Company No. 1 received a new Mack 1,250-gallon-per-minute pumper in 1970. In this view are Chief Mike Fitzgerald (second from left), Mayor Richard Blackstone (third from left), Assistant Chief Tony Cipolla (fourth from left), and master mechanic Edward Heimer (fifth from left) inspecting the truck. The other people in the picture are unidentified. (Courtesy of the East Hartford Fire Department.)

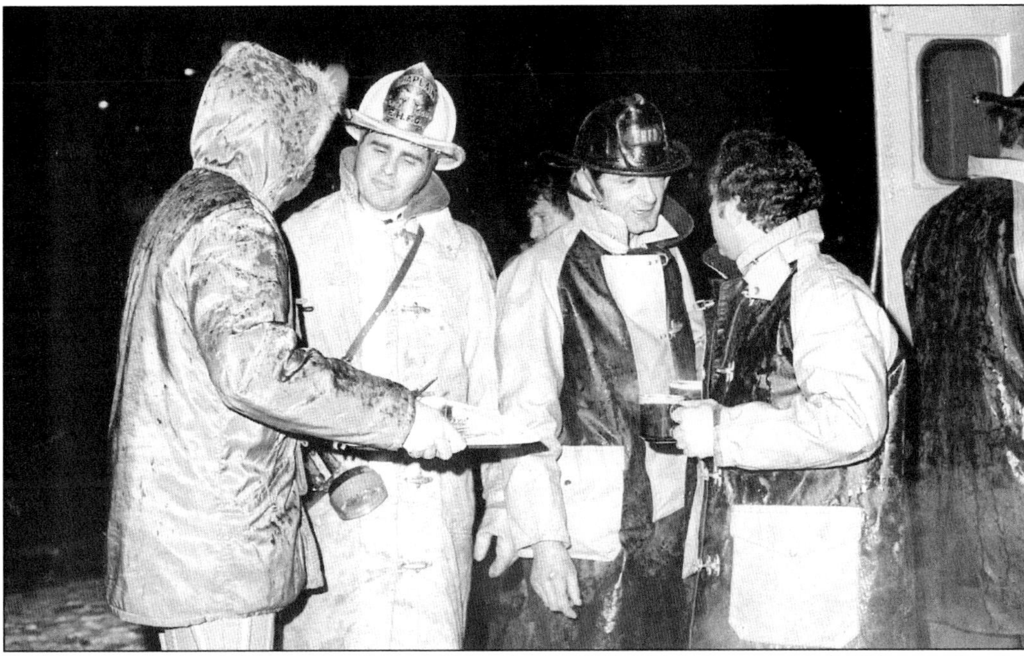

East Hartford Fire Department chaplain Stephen Foley is shown at this 1970s fire scene with firefighters Tony Rinaldi and Charlie Toce. Fire departments have had someone in the role of chaplain since the fire service was organized in the United States in the early 18th century. (Courtesy of the East Hartford Fire Department.)

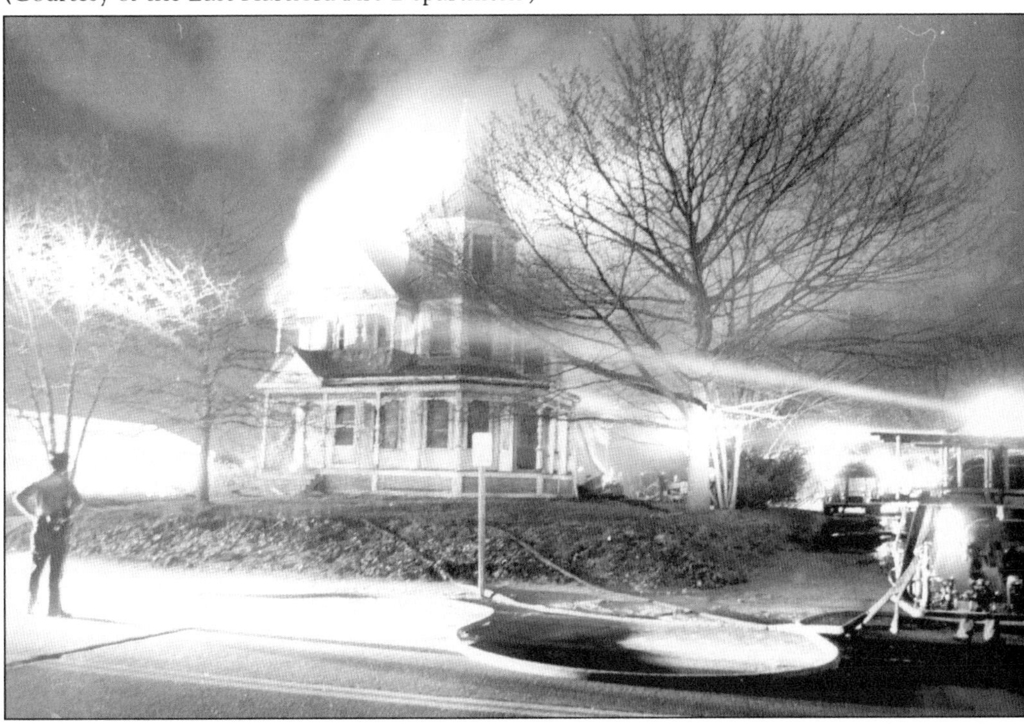

Fire is coming out of the upper story windows in this view of a Victorian house that caught on fire in this 1970s scene. (Courtesy of the Hartford Courant.)

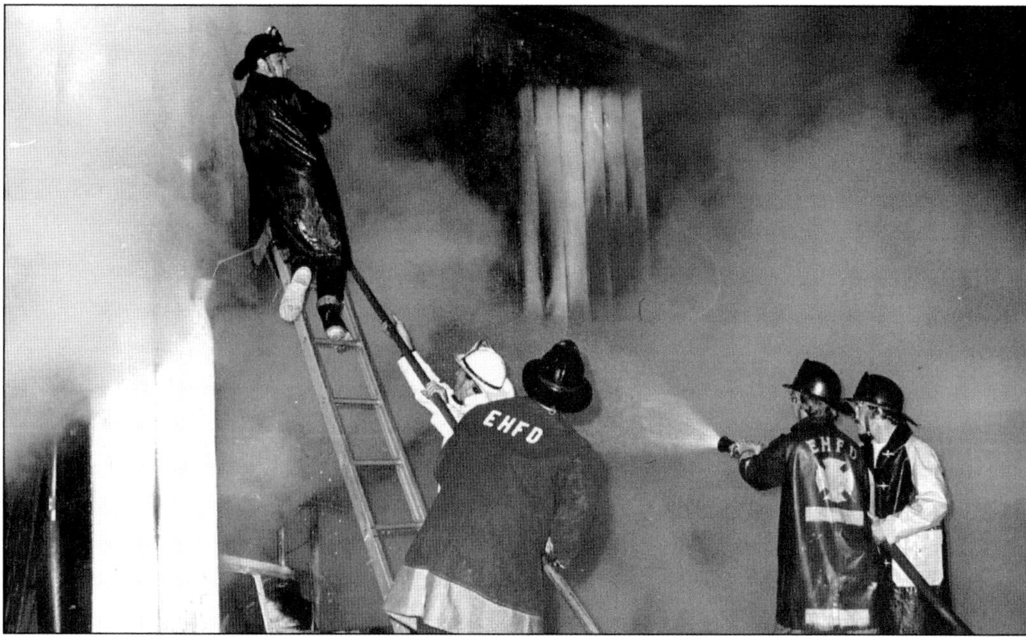

In the days before full protective gear, East Hartford firefighters are manning a line to help extinguish a barn fire in the 1960s. (Courtesy of the East Hartford Fire Department.)

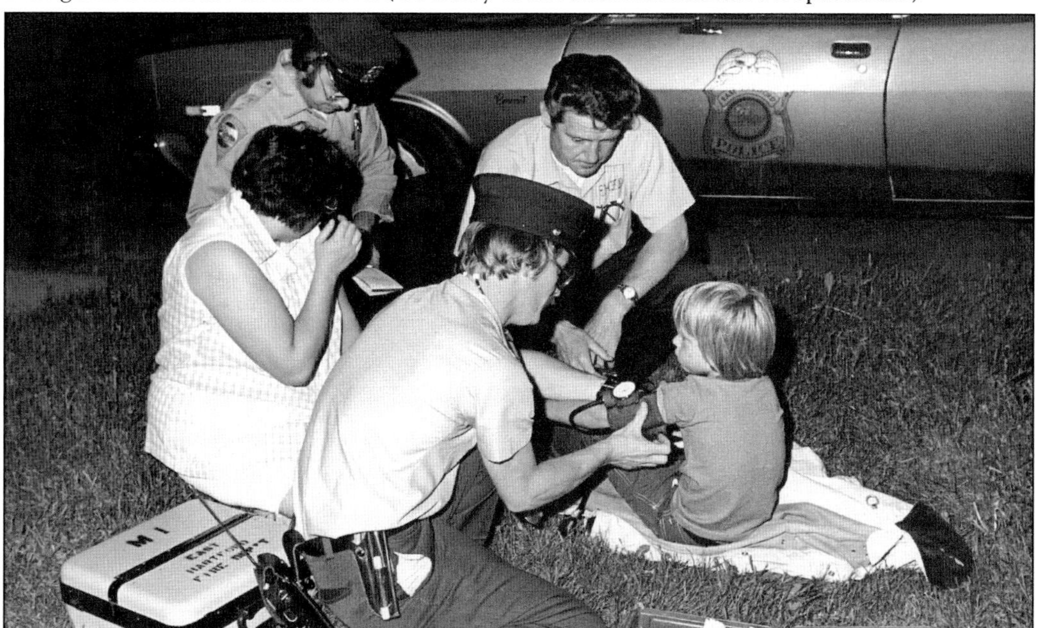

Firefighters Mike Coppinger (facing) and Dennis Garrity are treating this young patient in the late 1970s. Prior to the 1970s, fire departments were concerned primarily with fire extinguishment, and fire apparatus carried only basic first aid equipment. In 1972, the television show *Emergency* became an instant hit portraying the operation of the Los Angeles County Fire Department Paramedic Service. This show has been credited with increasing the need for EMS service to the community by the fire service. East Hartford was one of the first departments in Connecticut to have a dedicated paramedic vehicle. (Courtesy of the East Hartford Fire Department.)

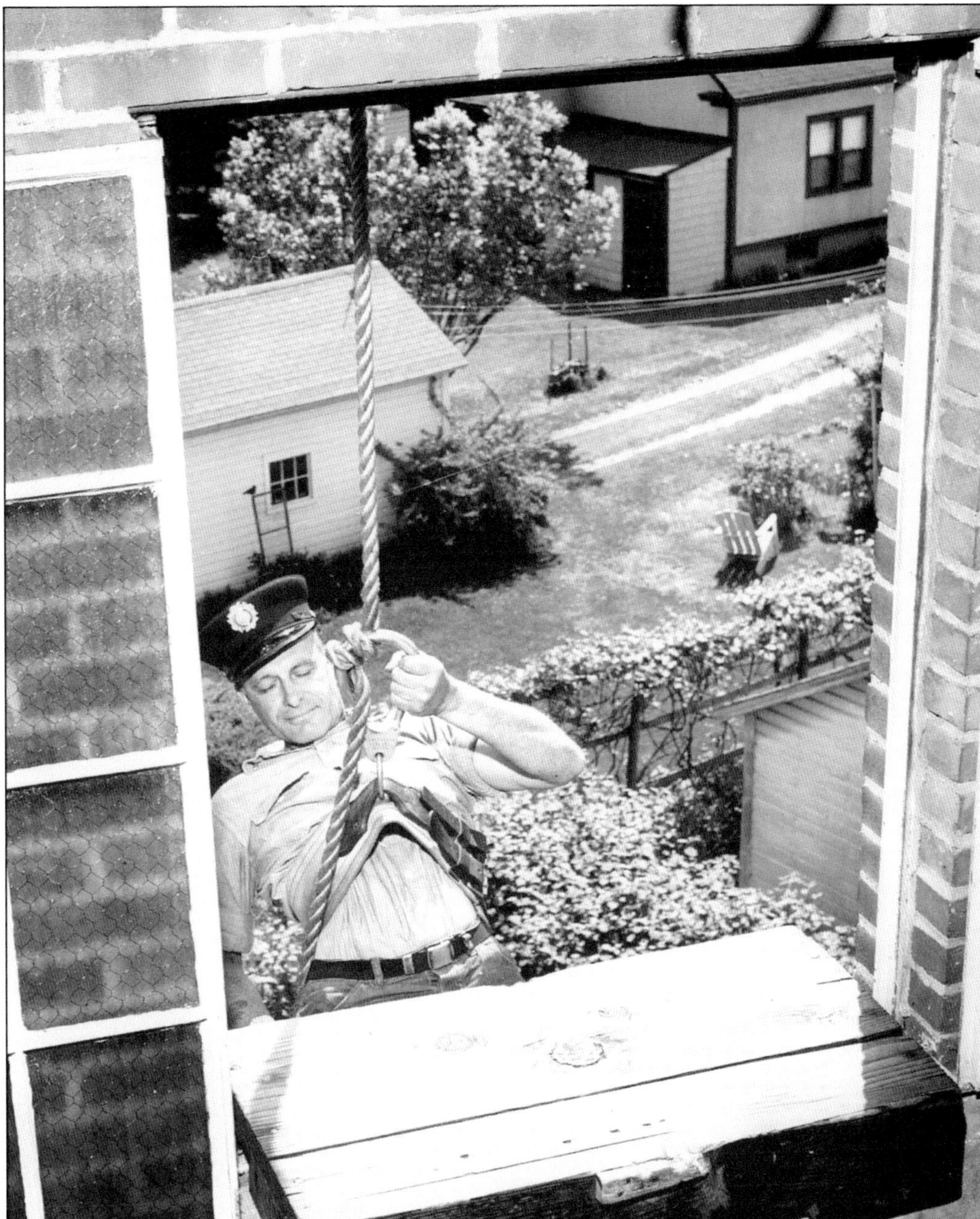

Firefighters for the past 100 years have undergone constant training with the gear and equipment they are required to use on the job. In this view is an East Hartford firefighter repelling down a rope from an upper story of a building, practicing a rescue that might be necessary in an extreme situation. All firefighters are required to learn how to use ropes and tie knots. (Courtesy of the East Hartford Fire Department.)

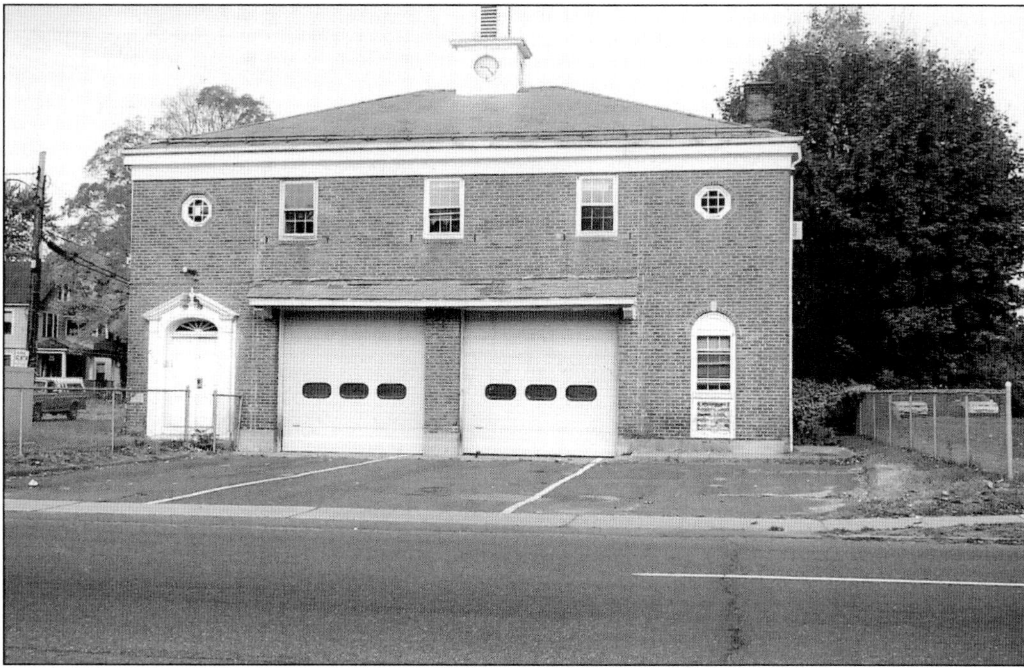

Seen here is a 2005 view of Station No. 3 after Engine No. 3 moved to its new quarters at 31 School Street. (Courtesy of the Connecticut Fire Museum Archives.)

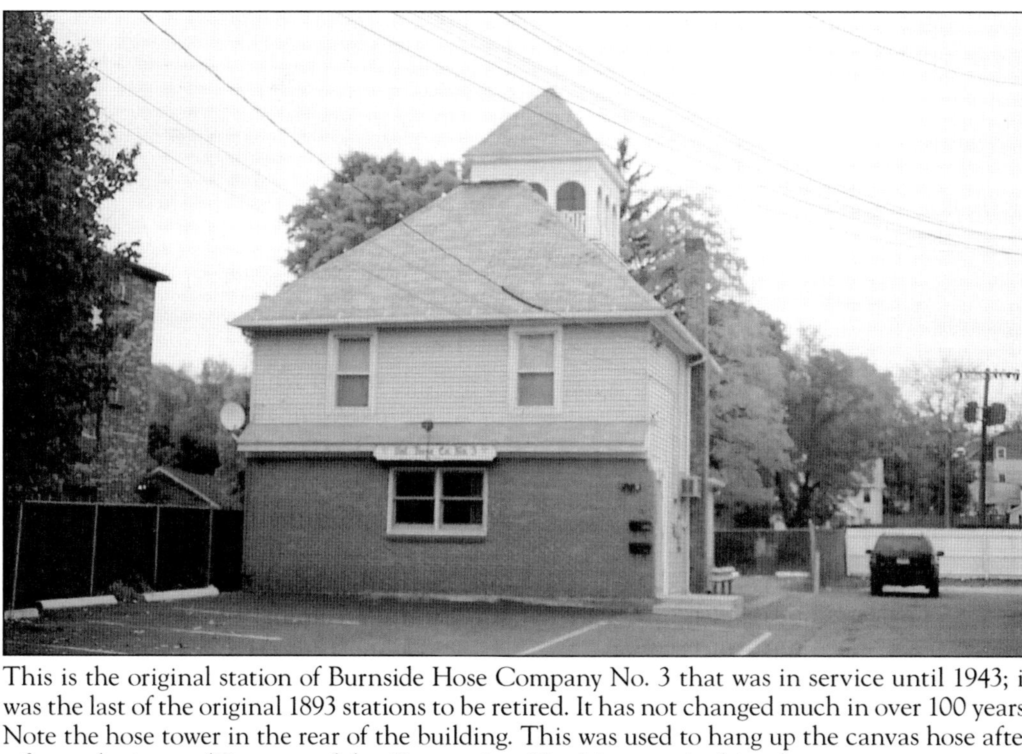

This is the original station of Burnside Hose Company No. 3 that was in service until 1943; it was the last of the original 1893 stations to be retired. It has not changed much in over 100 years. Note the hose tower in the rear of the building. This was used to hang up the canvas hose after a fire to dry it out. (Courtesy of the Connecticut Fire Museum Archives.)

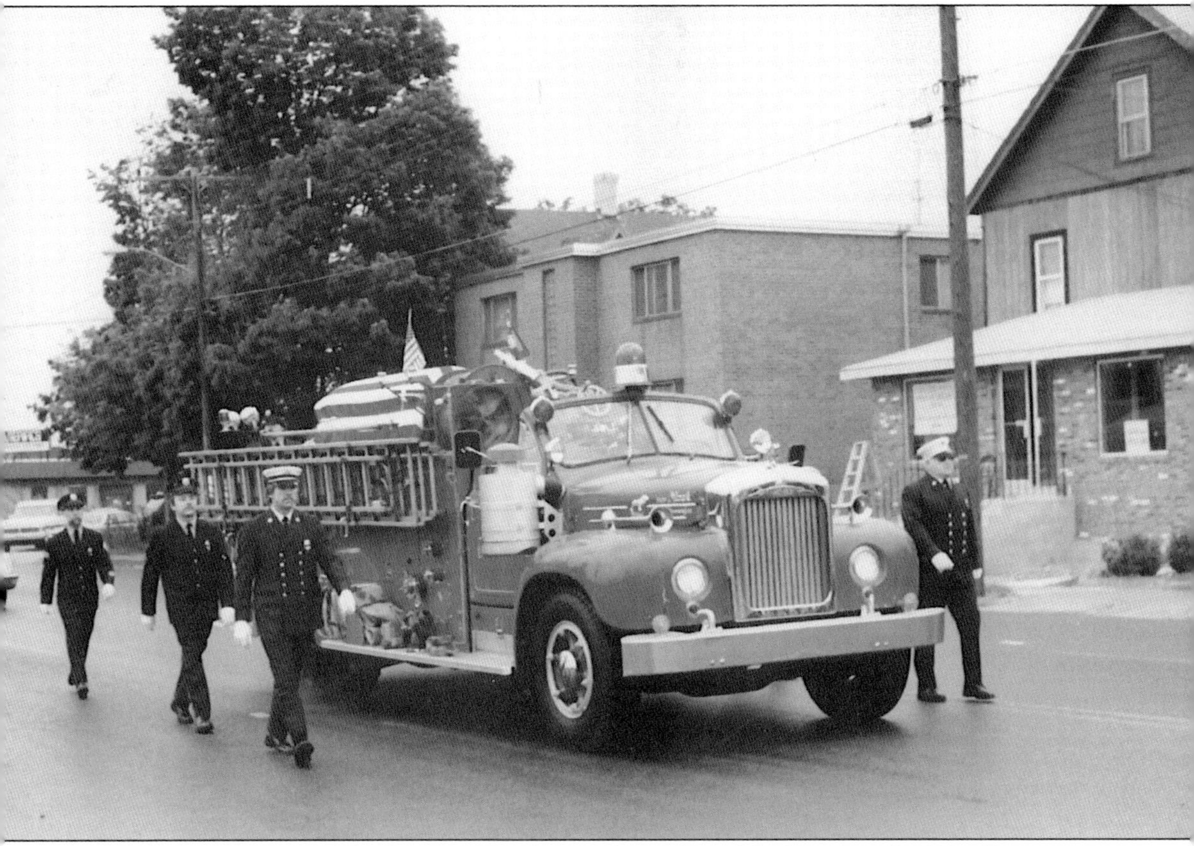

In the finest tradition of the fire service, a departed member of the East Hartford Fire Department is transported to the cemetery on Engine No. 3's apparatus as a tribute to his service to the community. (Courtesy of the Ken Beliveau Collection.)

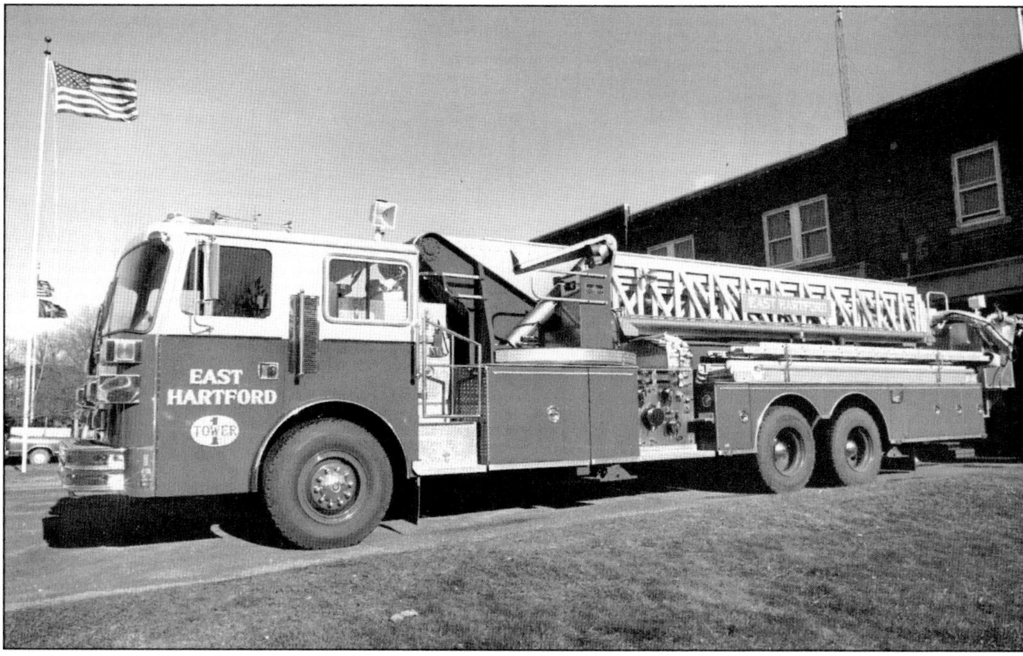

Ladder No. 1 operates this 1989 Sutphan tower ladder. This is classified as a "quint" in the fire service, as it is both a ladder truck and a pumper. It currently operates out of Station No. 3 on School Street. This unit replaced the 1968 American LaFrance tillered ladder truck. (Courtesy of the East Hartford Fire Department.)

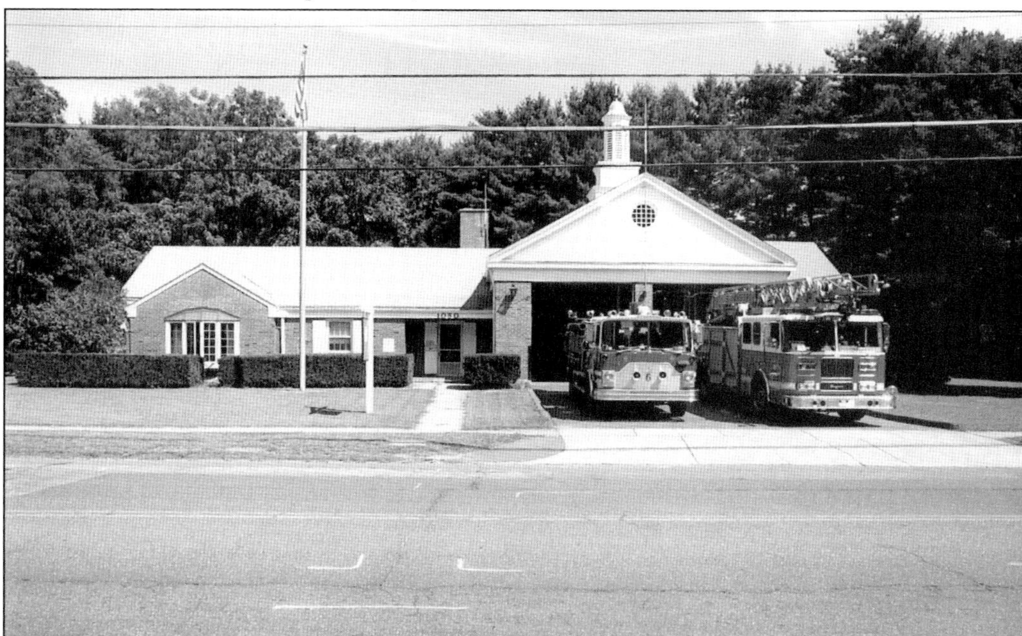

Station No. 6 is shown in this 2004 view. It was built in 1963 in response to the need for fire protection in the southeast section of town, which had seen substantial growth in the late 1950s. It is shown with Engine No. 6's Mack pumper and Ladder No. 2's Seagrave aerial ladder truck. Engine No. 6 received a new Seagrave pumper in 2006. (Courtesy of the Ken Beliveau Collection.)

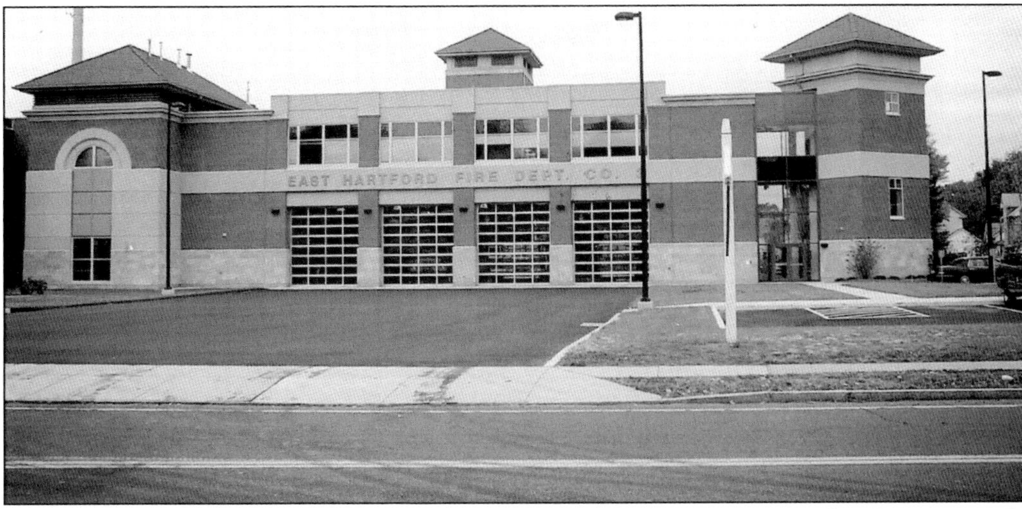

East Hartford's new Station No. 3 opened in 2003. At the same time, the department fire headquarters moved into the building from Station No. 1 on Main Street. All three buildings that served Engine Company No. 3 since 1893 still exist in 2006. (Courtesy of the Connecticut Fire Museum Archives.)

East Hartford Fire Chiefs

Joseph Taft	**1891-1894**
Edward O. Goodwin	**1894-1896**
Timothy J. Leary	**1897**
Edward Bragg	**1898**
Marshall Moore	**1900**
Edward O. Goodwin	**1901-1906**
Edward Bragg	**1906-1921**
Frank Meunier	**1921-1945**
John J. Armstrong	**1945-1949**
Francis J. Dagon	**1949-1963**
Dennis S. McCarthy	**1963-1967**
John J. Kelly	**1967-1969**
Michael J. Fitzgerald	**1969-1980**
Thomas W. Dawson, Jr.	**1980-1990**
David Dagon	**1991-2003**
Michael Eremita	**2003 - Present**

This is a list of the East Hartford fire chiefs from 1891 to 2006. (Courtesy of the Connecticut Fire Museum Archives.)

Stations and Apparatus

1925

Engine Co. #1	**1917 American LaFrance 759 gpm pumper hose and chemical car.**
Bissell Street	
Engine Co. #2	**1921 REO Chemical and Hose Car**
Park Avenue and Main Street	
Engine Co. #3	**1923 Mack 350 gpm pumper and hose car**
12 Hanmer Street	

1938

Engine Co. #1	**1929 Mack 1000 gpm pumper**
726 Main Street	
Engine Co. #2	**1934 Mack 600 gpm pumper**
Park Avenue and Main Street	
Engine Co. #3	**1923 Mack 350 gpm pumper**
12 Hanmer Street	
Engine Co. #5	**1917 American LaFrance 750 gpm pumper**
304 Main Street	

1956

Engine Co. #1	**1953 Mack 1000 gpm pumper**
Squad "A"	**1955 Mack 500 gpm pumper**
Ladder Co. #1	**1948 American LaFrance 75' aerial**
726 Main Street	
Engine Co. #2	**1934 Mack 600 gpm pumper**
1692 Main Street	
Engine Co. #3	**1940 Mack 500 gpm pumper**
621 Burnside Avenue	
Engine Co. #5	**1940 Mack 500 gpm pumper**
Foam Unit	**1941 Chevrolet Foam truck.**
304 Main Street	

This image displays East Hartford fire stations and apparatus from 1925 to 1956. (Courtesy of the Connecticut Fire Museum Archives.)

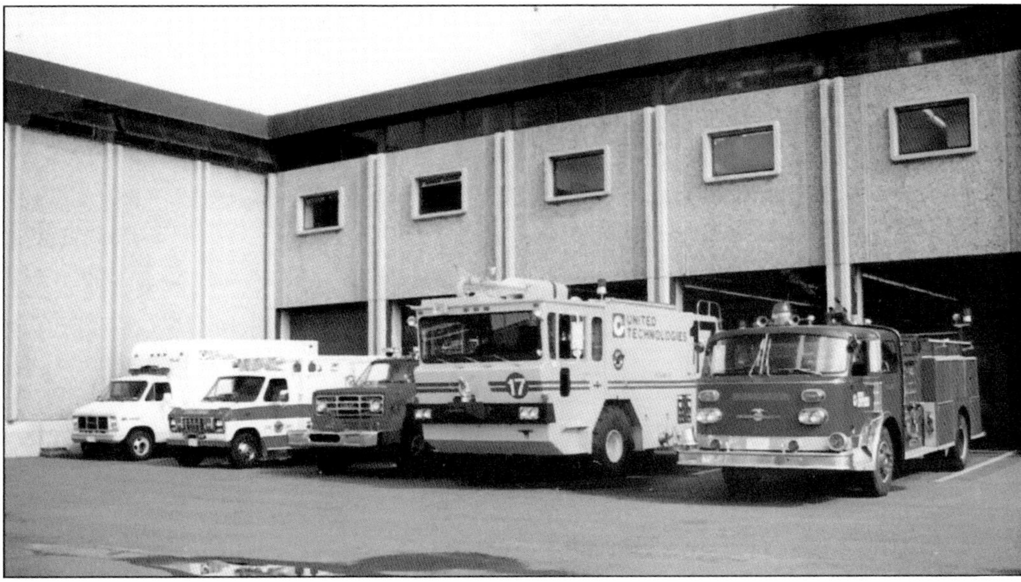

The largest private fire department in Connecticut is located at the East Hartford plant of Pratt and Whitney Aircraft Division of United Technologies Corporation; located on 1,080 acres, the manufacturing complex consists of over 2 million square feet of factory space, bulk fuel storage, and, until 1998, the second-largest airport in Connecticut, Rentschler Field. In this view is the Pratt and Whitney Aircraft fire station located in the middle of the complex along with the department equipment that includes an airport crash truck. The Pratt and Whitney Fire Department and the East Hartford Fire Department have always had a close working relationship with Pratt and Whitney Aircraft apparatus responding with its specialized fire suppression equipment, especially with petroleum fires, when needed to assist. (Courtesy of the Connecticut Fire Museum Archives.)

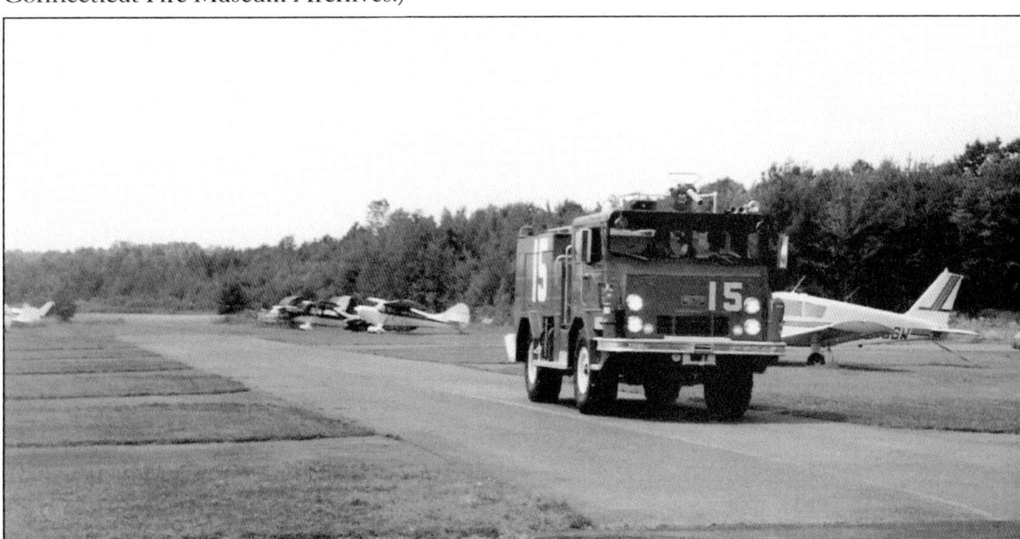

A 1967 Walter airport crash truck operated as a light rescue vehicle by the Pratt and Whitney Aircraft Fire Department at their East Hartford plant. It carried 2,000 pounds of dry chemical for aircraft-type fires. Retired in 1996, this vehicle is now on display at the Connecticut Fire Museum in East Windsor. (Courtesy of the Connecticut Fire Museum Archives.)

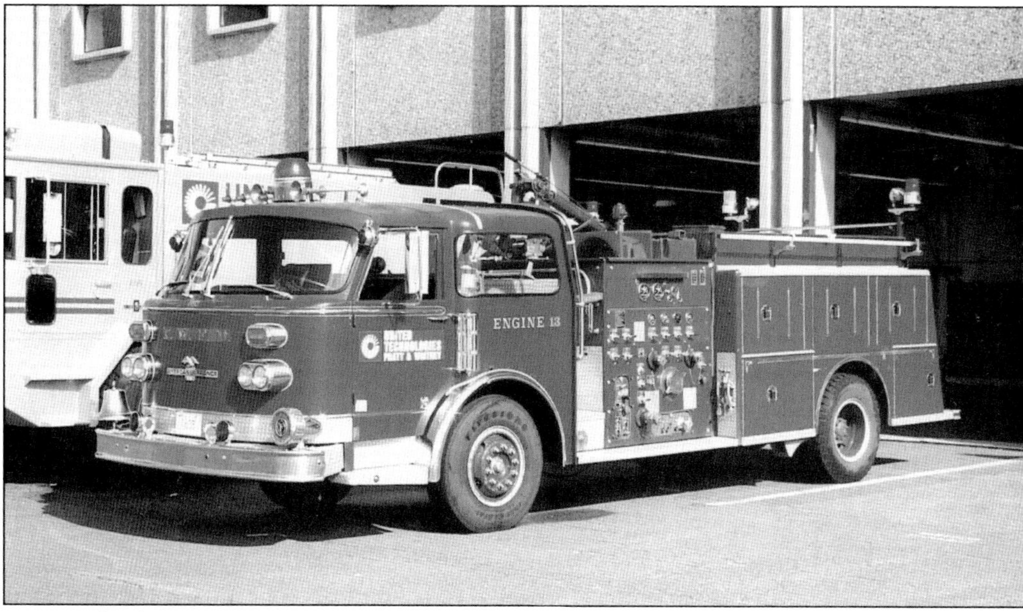

Pratt and Whitney Aircraft Fire Department's Engine No. 13 is this 1983 American LaFrance 1,000-gallon-per-minute pumper that formally served at Sikorski Aircraft in Stratford, Connecticut, before coming to East Hartford. The Pratt and Whitney Aircraft department responds to over 1,000 incidents per year, all within this vast factory complex with its staff of 50 career firefighters. (Courtesy of the Connecticut Fire Museum Archives.)

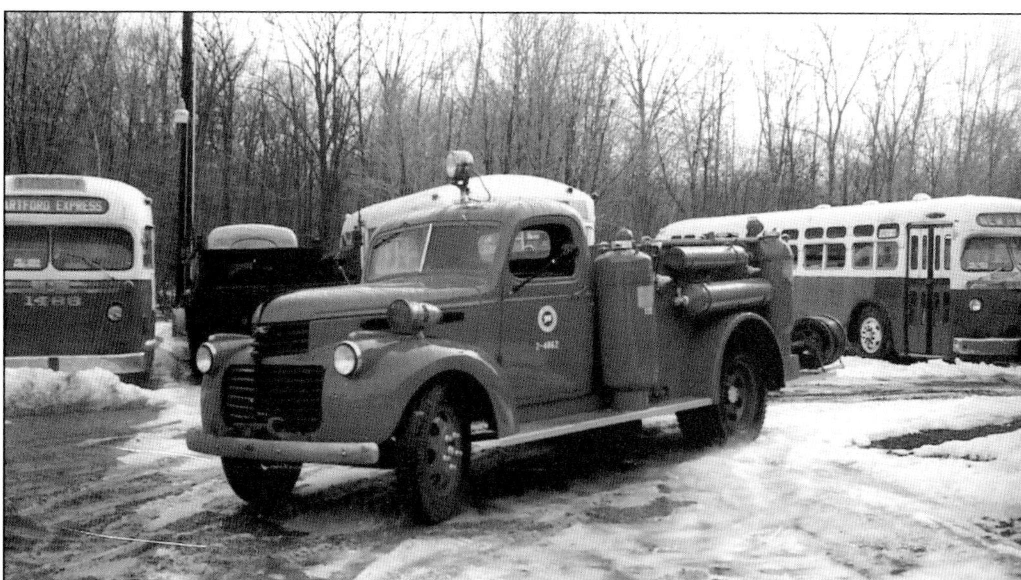

In 1943, Pratt and Whitney Aircraft purchased this small airport crash truck, built on a GMC chassis by the Sealand Corporation of Bridgeport, Connecticut; this unit was one of the early airport crash trucks built in the United States. In 1985, this truck was donated to the State of Connecticut for use at Brainard Airport in Hartford where it operated until 1990. It is now on display at the Connecticut Fire Museum in East Windsor. It carries 300 pounds of dry chemical. (Courtesy of the Connecticut Fire Museum Archives.)

Six

West Hartford Fire Department

Initially, the town of West Hartford had five independent fire companies. Three were incorporated fire districts that operated with paid firefighters, and two were operated with volunteers. West Hartford Hose Company was organized first on July 24, 1879. Their equipment consisted of 100 feet of obsolete leather hose, a hydrant wrench, and one play pipe nozzle. The first hydrant was installed in the center of town in October 1879. The honor of turning on the hydrant was given to Chief Henry J. Eaton of the Hartford Fire Department, who was also an honorary member of the West Hartford Hose Company and lived on Mountain Road. On January 19, 1885, the organization name was changed to the Fountain Hose Company, and John E. Millard was elected chief. Over the years, more modern canvas hose along with a two-wheel hose cart were obtained. In February 1915, a secondhand Pope-Hartford truck was purchased and outfitted for fire service with a water tank and all the necessary hose, nozzles, and tools. The company also had a 1917 Pierce-Arrow hose car. The Pope-Hartford truck was operated until a new Stevens-Duryea truck was obtained in 1921, when the Pope-Hartford apparatus was disposed of.

The first movement to have fire protection in the eastern portion of West Hartford was in 1909, when a meeting was held at the Whiting Lane School on November 22, 1909, when the East Side Fire District was organized. The boundaries of the district consisted of all the area bounded by Park Road, Whiting Lane, Steele Road, Prospect Avenue, and Albany Avenue. In 1912, a lot was purchased on Prospect Avenue and the fire station, which stands today, was erected in 1915. A 1915 American LaFrance pumper and hose car along with five firefighters were housed there. An agreement was signed with the Hartford Fire Department for mutual aid to have the East Side Fire District respond on the first alarm to eight Hartford fireboxes located on Prospect Avenue. The company would also cover for Engine No. 12, on South Whitney Street, when they were on a call. The agreement called for Hartford to respond as necessary to fires in the district and cover the district in case of a second fire. The equipment at the station was compatible with Hartford fire apparatus. In 1916, the Fountain Hose Company No. 1 offered to serve as a covering company for Hartford in case of an emergency. In 1926, an American LaFrance pumper-ladder truck was added to the Prospect Avenue Station, which now had 10 firefighters on duty. In 1930, the fire district was expanded when the East Side Fire District opened a new station on Bloomfield Avenue. This new station was added to the mutual aid agreement between the district and Hartford. The station was equipped with a new 1930 American LaFrance 200 series pumper similar to four that the Hartford Fire Department had also purchased in 1930.

The Center Fire District was organized at a meeting held on May 4, 1914. For several years, the district shared the quarters of Fountain Hose Company in the rear of the Baptist Church. In 1925, the district built a new station on Brace Road. A 1915 American LaFrance pumper was transferred from the Prospect Avenue station when a new American LaFrance pumper replaced it. In 1926, a Maxim cities service ladder truck was put in service. The Quaker Hose Company was organized on January 1, 1914, and operated from a small station on Boulevard that resembled a small house. On May 8, 1919, the South Side Fire District was organized and only operated within a rigidly defined fire protection area, not providing mutual aid to other areas of the town. In 1920, the Elmwood Hose Company was organized, but by November 1921, it was merged in the South Side District. In November 1921, the South Side District received delivery of a Maxim triple combination pumper; this was the first of many pieces of Maxim apparatus West Hartford would purchase over the next 50 years. In 1930, the district opened a new station on New Britain Avenue, which remains in service in 2006.

On March 1, 1937, the town of West Hartford merged the independent fire districts into one unified municipal department. At the time of consolidation, Fountain Hose Company was disbanded. The Quaker Hose Company was asked to maintain two callmen to operate a floodlight truck, which responded to all fire alarms at night. They continued in this capacity until being disbanded in 1968. In 2006, the West Hartford Fire Department consists of five stations, 93 career firefighters, and staff represented by Local 1241 of the International Association of Firefighters, headed by Chief William Austin. The department responds to about 5,000 fire and medical calls a year.

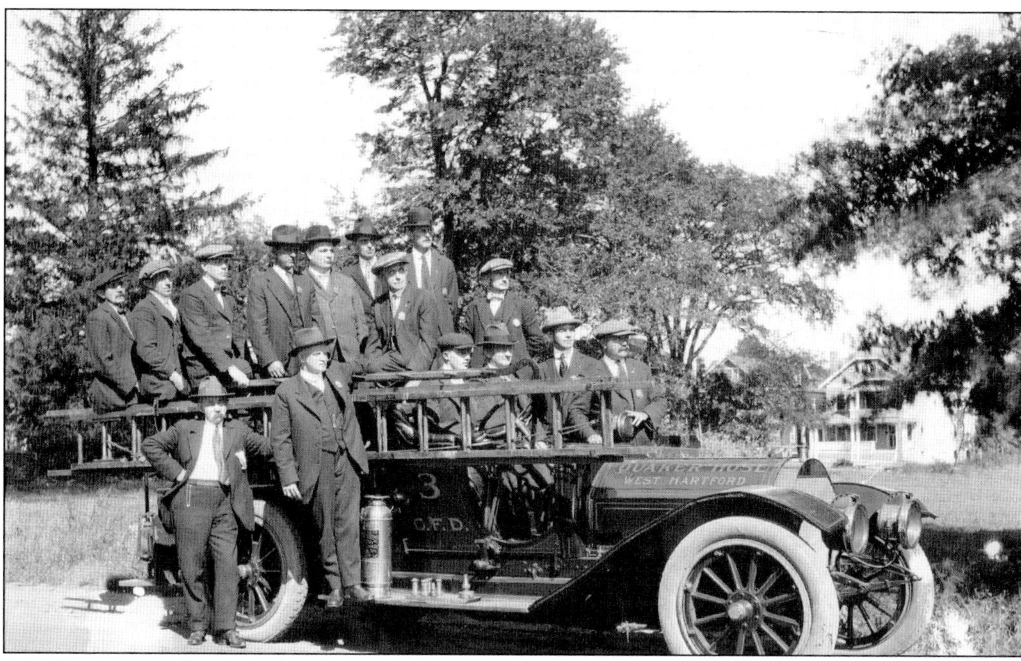

Seen here is a photograph of Quaker Hose Company members and their 1919 Pierce-Arrow hose car. In 1914, Quaker Hose used a rebuilt delivery wagon, hand drawn, with 500 feet of secondhand hose and 25 leather buckets. The hand-drawn unit was soon replaced by a Mitchell touring car rebuilt into a hose wagon. In 1919, the Mitchell was replaced by the Pierce-Arrow. Most of money was obtained by running clambakes. In gratitude to the citizens of the area for their generosity, the Quaker Hose Company made a practice of hauling stalled cars out of the mud. (Courtesy of the Rudy Desroches Collection.)

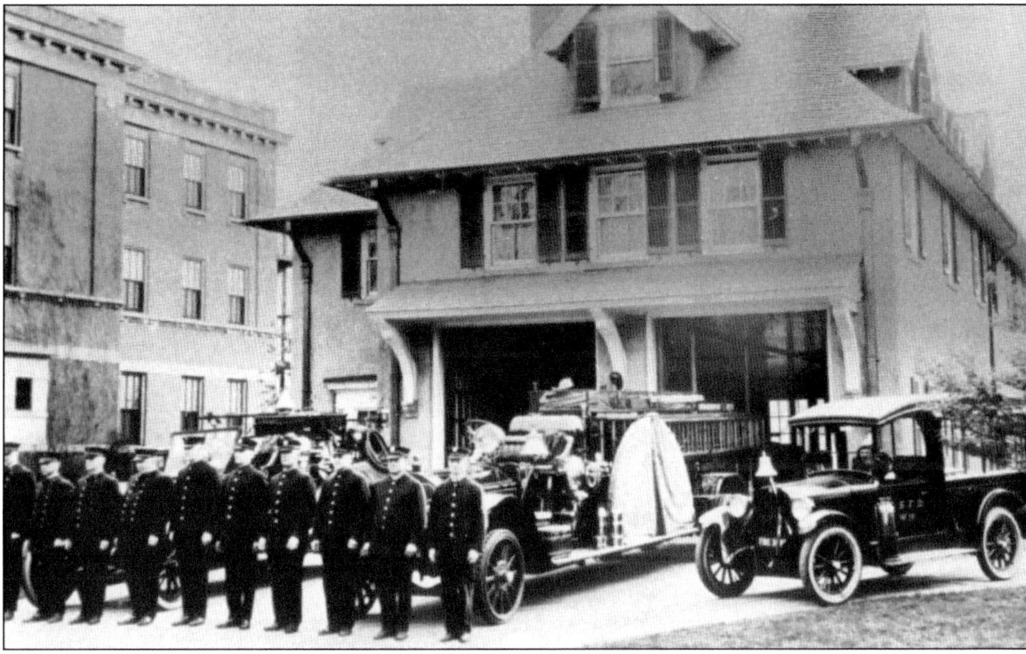

Firefighters of the East Side Fire District of West Hartford stand at attention with their apparatus at the Prospect Avenue station around 1923. In the background is the apparatus that operated from this station: a 1915 American LaFrance pumper and hose car, and a 1920 American LaFrance combination pumper-ladder truck called a "quad" in the fire service. This station was built in 1915 as part of the mutual aid agreement between Hartford and the East Side Fire District. This station remains in service in 2006 as West Hartford Fire Department Station No. 1. (Courtesy of the Rudy Desroches Collection.)

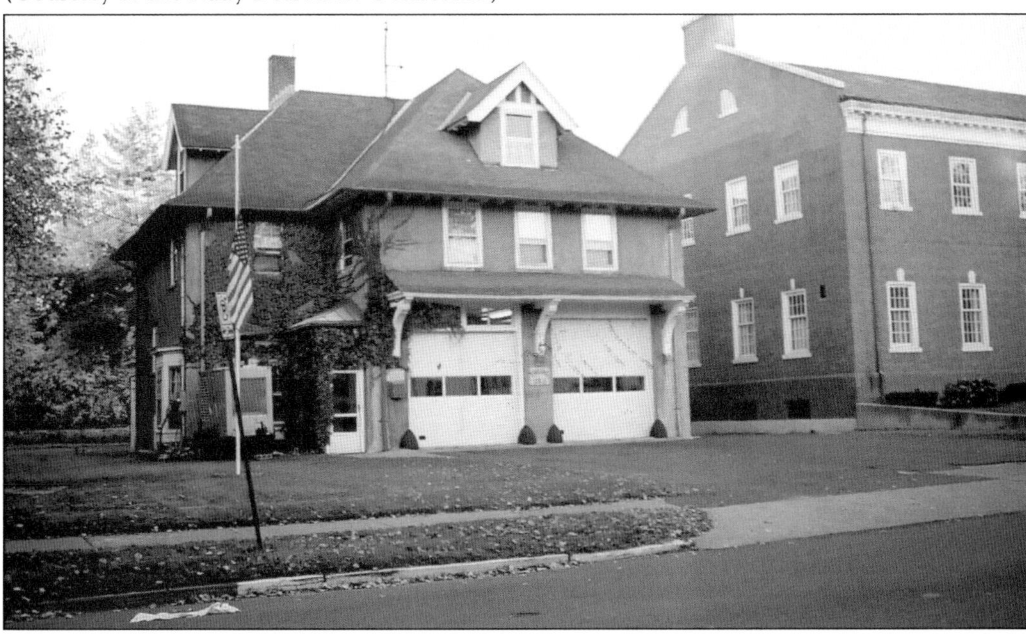

West Hartford Fire Department Station No. 1 is shown in this 2005 view. (Courtesy of the Connecticut Fire Museum Archives.)

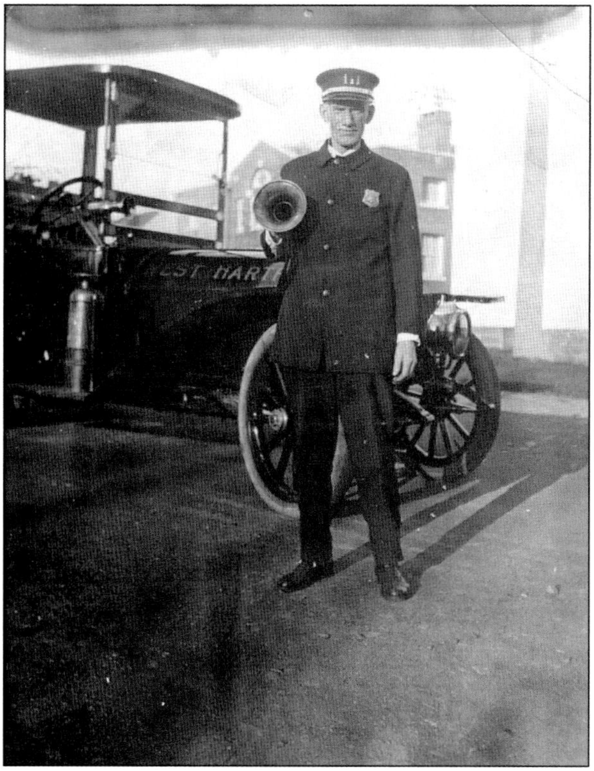

Chief James Livingston of Fountain Hose Company is seen standing next to one of the company hose wagons around 1917. Not all sections of town were protected by the three regular fire districts, thus the need for the volunteer Quaker Hose and Fountain Hose companies. The cost of fire protection varied in each district but added only three or four mills to the property tax bills for property owners. (Courtesy of the Rudy Desroches Collection.)

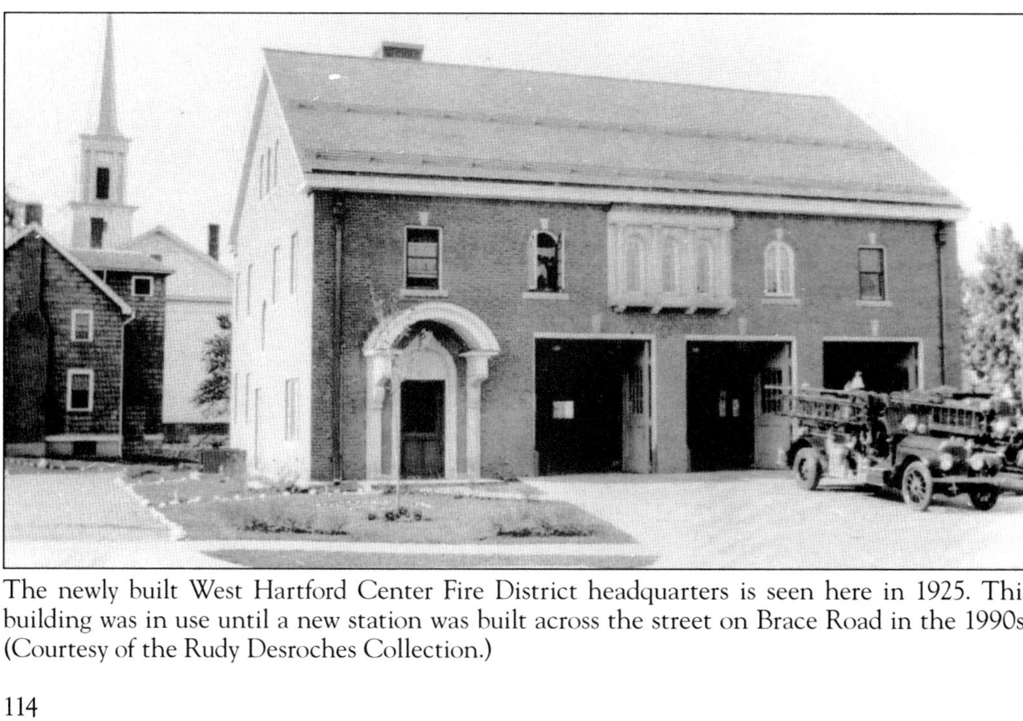

The newly built West Hartford Center Fire District headquarters is seen here in 1925. This building was in use until a new station was built across the street on Brace Road in the 1990s. (Courtesy of the Rudy Desroches Collection.)

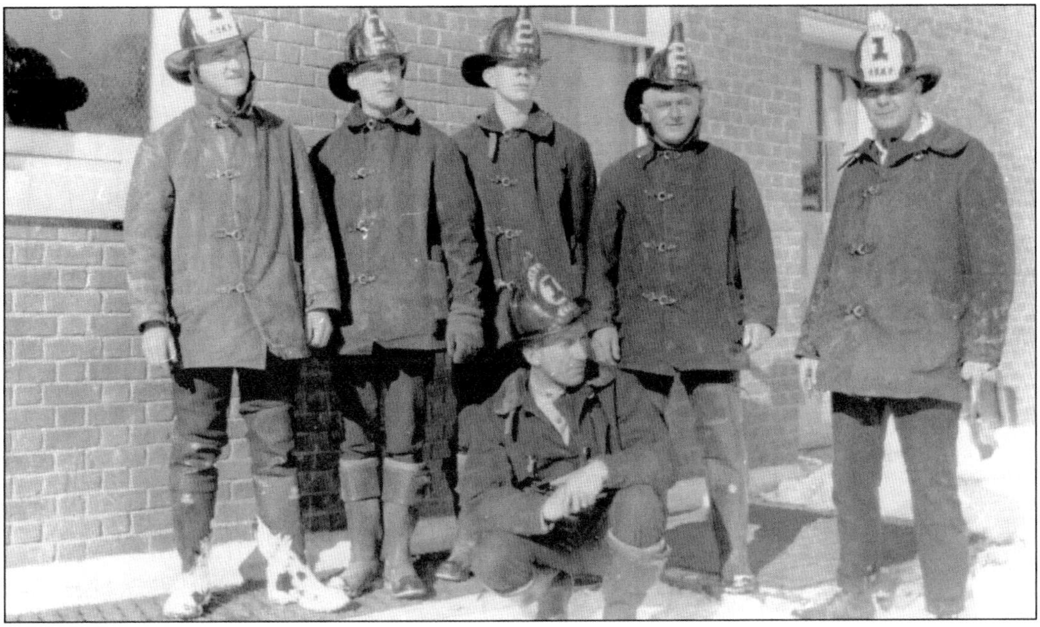

Center Fire District firemen at Station No. 2 are, from left to right, Nicholas Angelo, Emil Mahlendorf, Chief Thomas Donnelly, William Dooley, and John T. O'Laughlin, who succeeded Donnelly as chief in 1956. The fireman kneeling in the front is unidentified. (Courtesy of the Rudy Desroches Collection.)

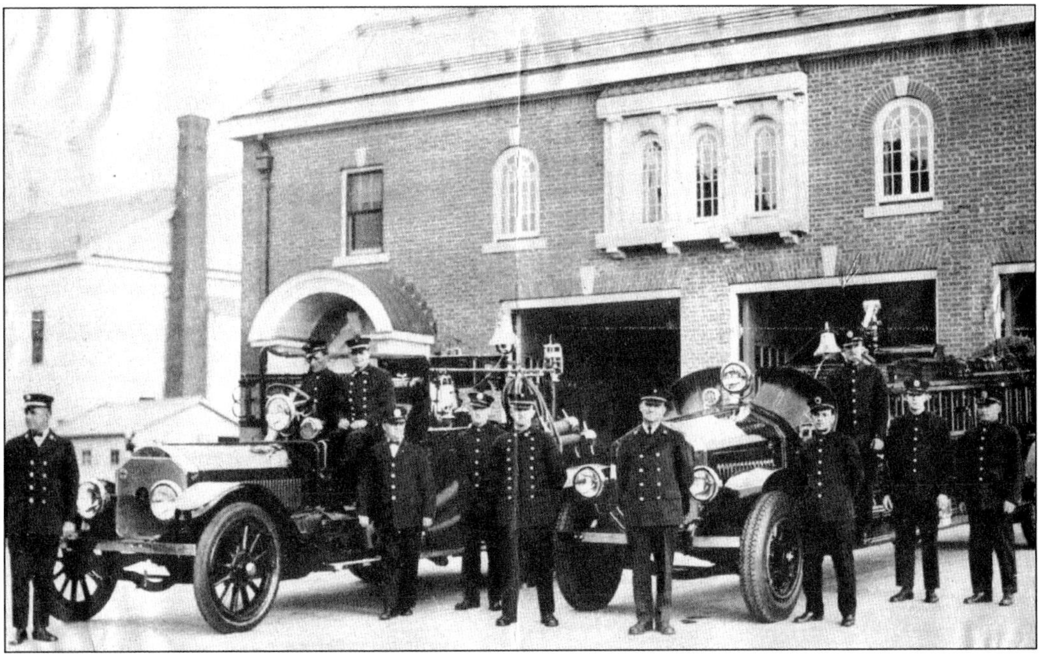

A 1915 American LaFrance pumper and hose car and a 1926 Maxim ladder truck of the West Hartford Center Fire District are shown in this 1927 view. Firefighters, shown left to right, are master mechanic E. L. Burdick, George Jones, Lt. E. J. Fox, J. H. Young, Walter Mills, Lt. K. T. Brumbaum, Capt. T. H. Donnelly, Frank Fidrino, C. H. Shaw, Russel Noyes, and P. J. Conners. (Courtesy of the Rudy Desroches Collection.)

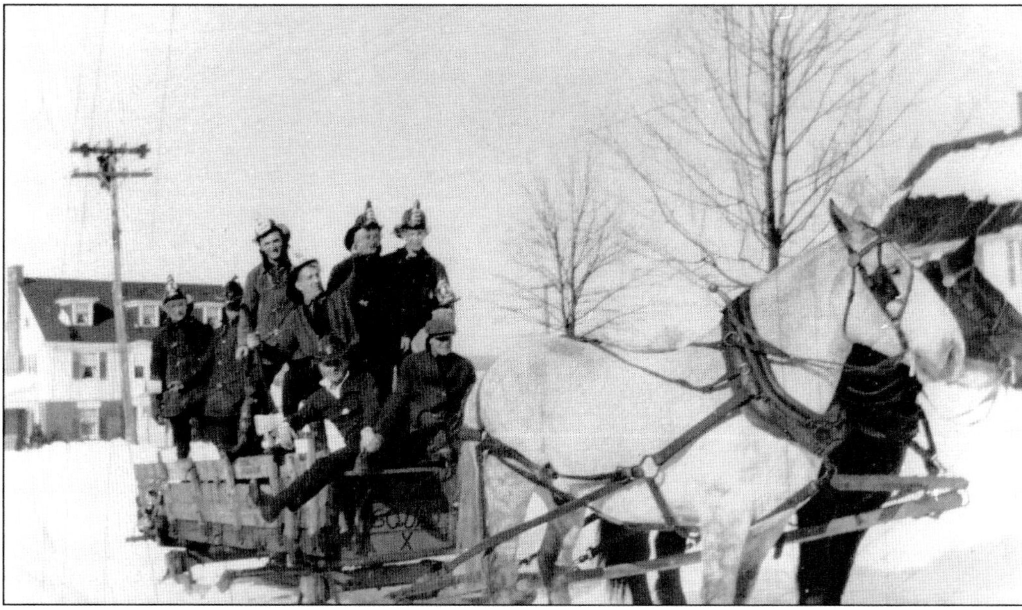

West Hartford firefighters use a horse-drawn hose sleigh in this 1925 photograph. Horse-drawn sleighs were very common sights in the wintertime before 1914, as only those streets that had streetcar lines on them were plowed after a storm in the winter and that was by the streetcar company. Side streets were many times impassable in the winter to the early apparatus, and before that the horse-drawn steamers and hose wagon, so sleighs were used by the firefighters. (Courtesy of the Rudy Desroches Collection.)

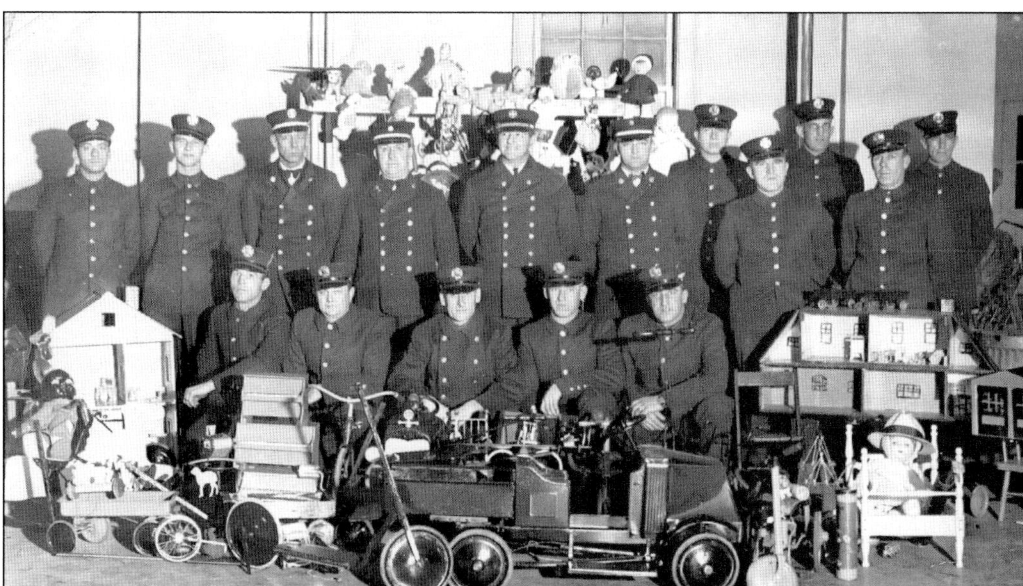

West Hartford firefighters are seen running a toy drive at Christmas at Station No. 3 in 1932. (Courtesy of the Hartford Courant, Rudy Desroches Collection.)

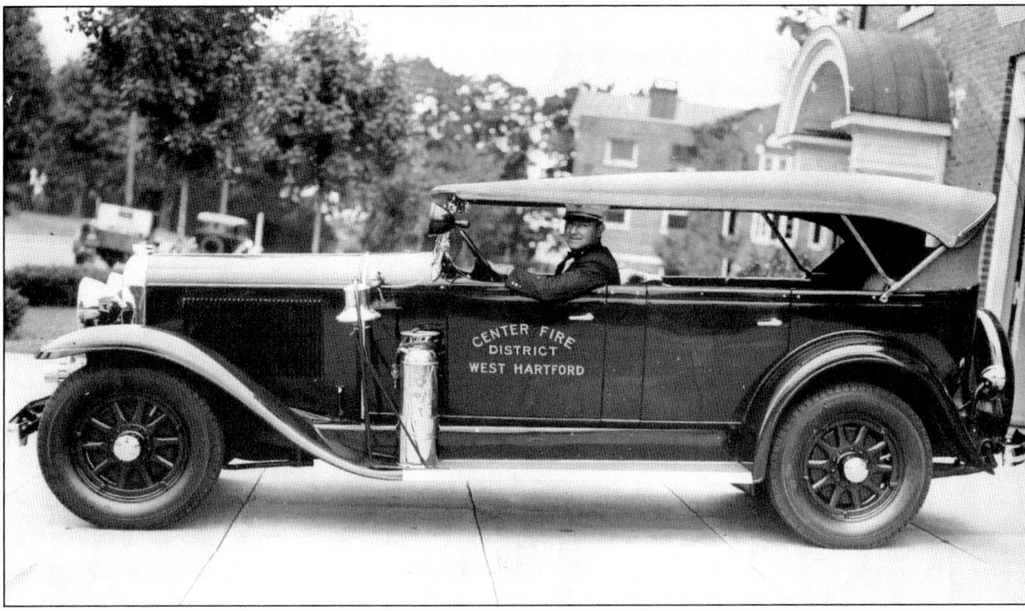

This chief rode to fires in style in this big touring car, shown in front of the Center Fire District headquarters in the 1930s. (Courtesy of the Rudy Desroches Collection.)

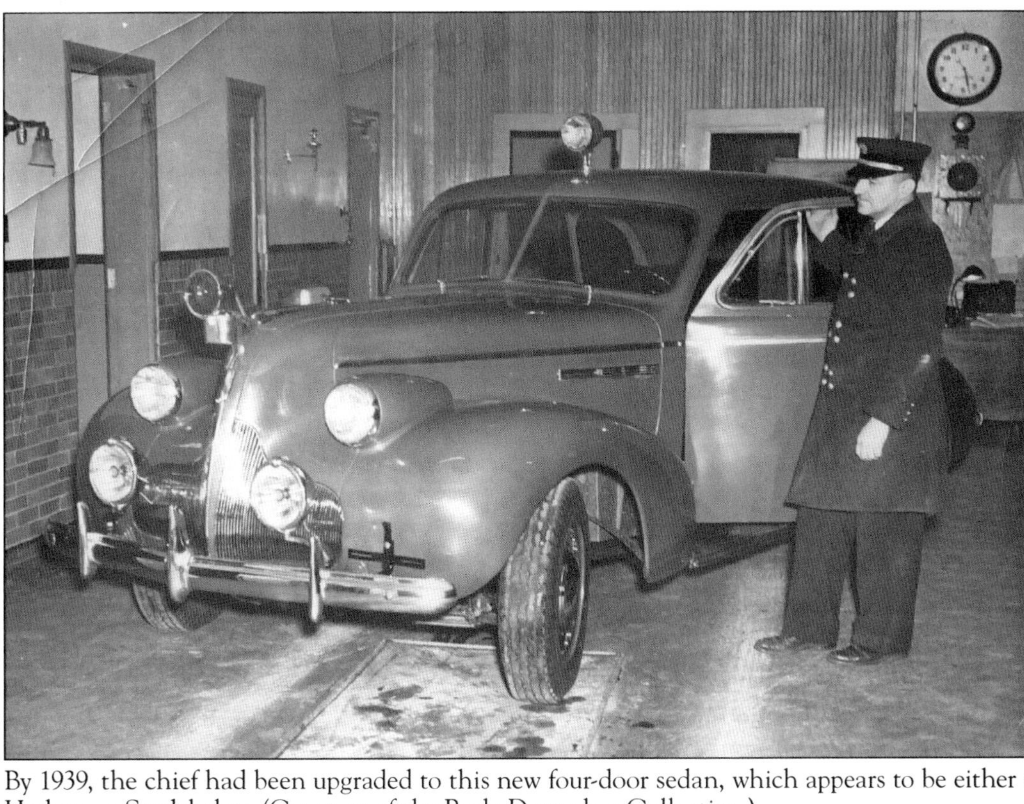

By 1939, the chief had been upgraded to this new four-door sedan, which appears to be either a Hudson or Studebaker. (Courtesy of the Rudy Desroches Collection.)

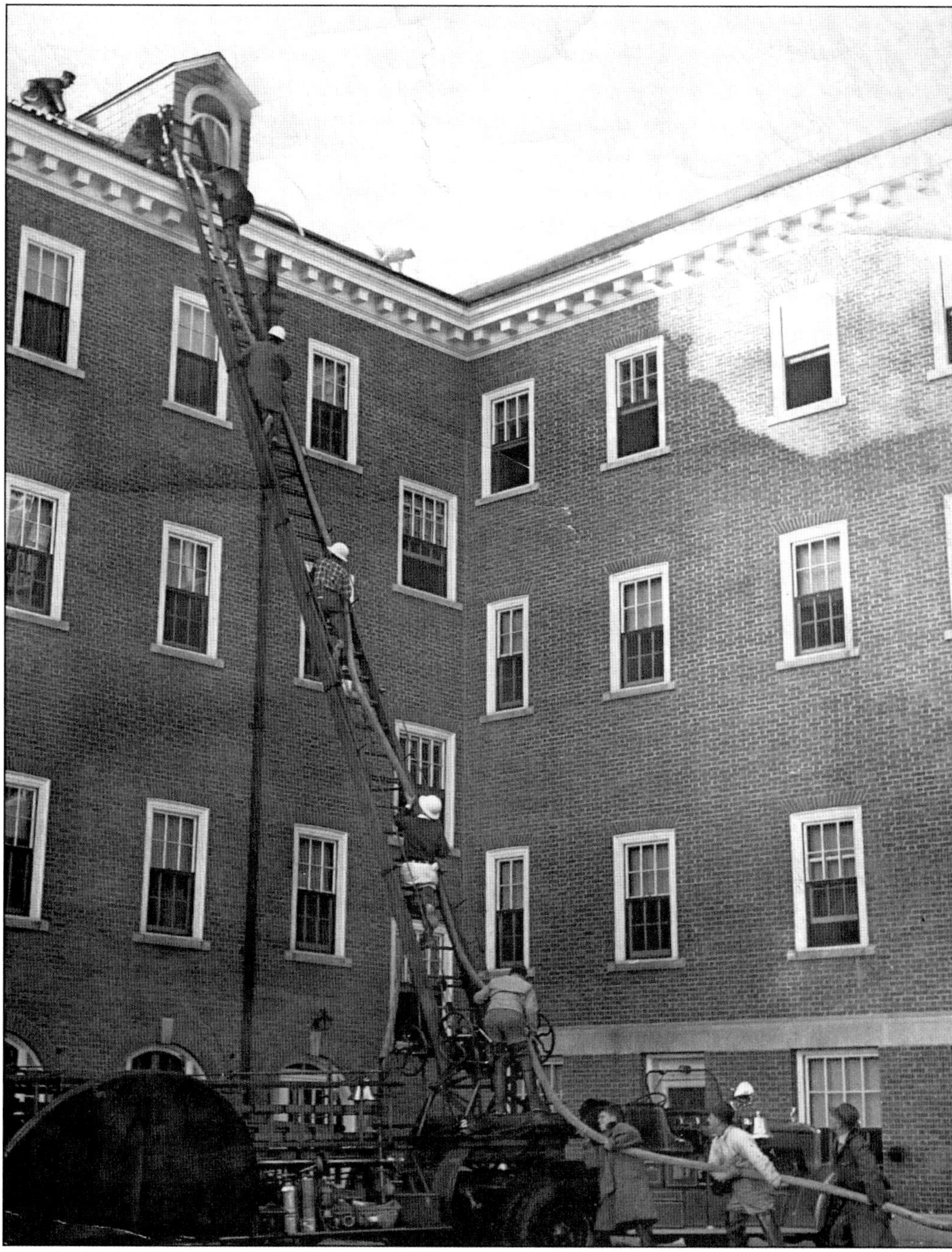

Hartford's Ladder No. 5 is shown in this scene operating at a fire at the Mary Immaculate Convent on Park Road on November 7, 1943. Climbing the ladder are some of the World War II–era auxiliary firemen who were deployed for this fire. The 1934 Mack tractor had a 1910 American LaFrance 85-foot aerial ladder trailer that was originally horse drawn. It was sold to a Hartford firefighter in the 1970s. Any fire along the city line had both Hartford and West Hartford responding as a result of the 1915 mutual aid agreement. (Courtesy of the Rudy Desroches Collection.)

In 1930, the East Side Fire District opened a new station on Bloomfield Avenue with a 1930 American LaFrance 200 series pumper similar to the four Hartford purchased the same year. After the 1937 reorganization of the fire departments, this became Company No. 4 of the West Hartford Fire Department. (Courtesy of the Rudy Desroches Collection.)

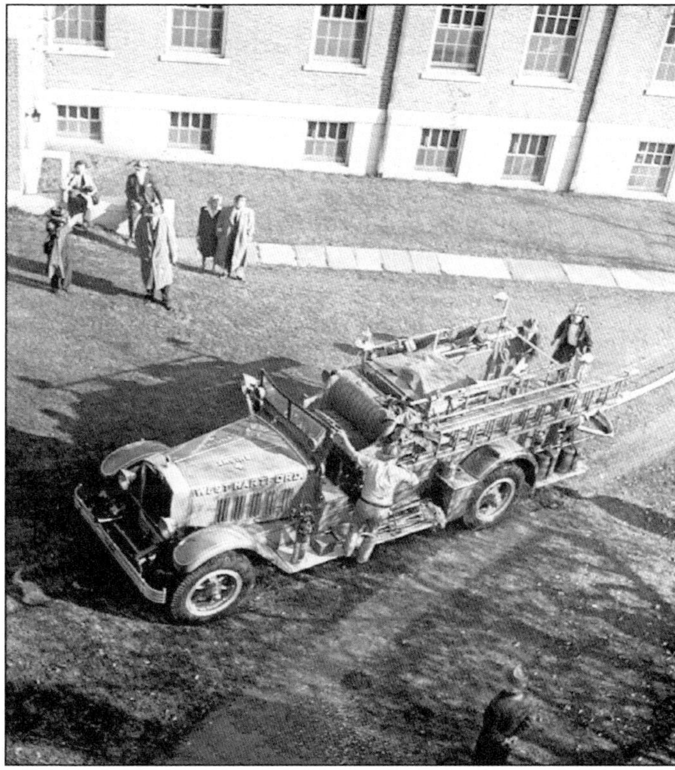

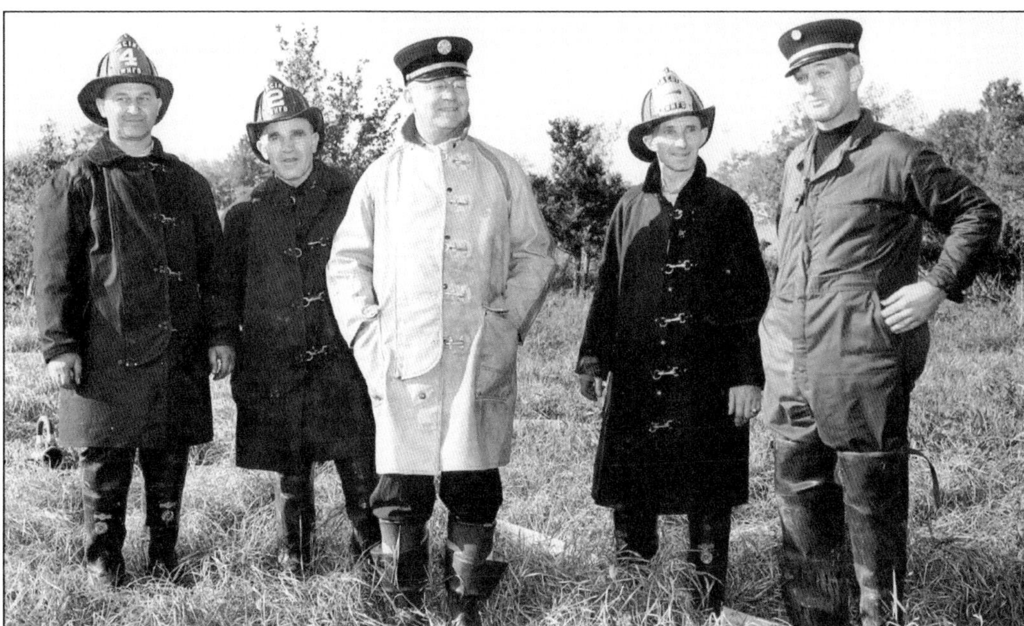

Firefighters and officers of the West Hartford Fire Department are, from left to right, Nicholas Angelo, Emil Mahlendorf, Chief Thomas Donnelly, William Dooley, and John T. O'Laughlin who succeeded Donnelly as chief in 1956. (Courtesy Rudy Desorches Collection, photograph by R. A. Stevens.)

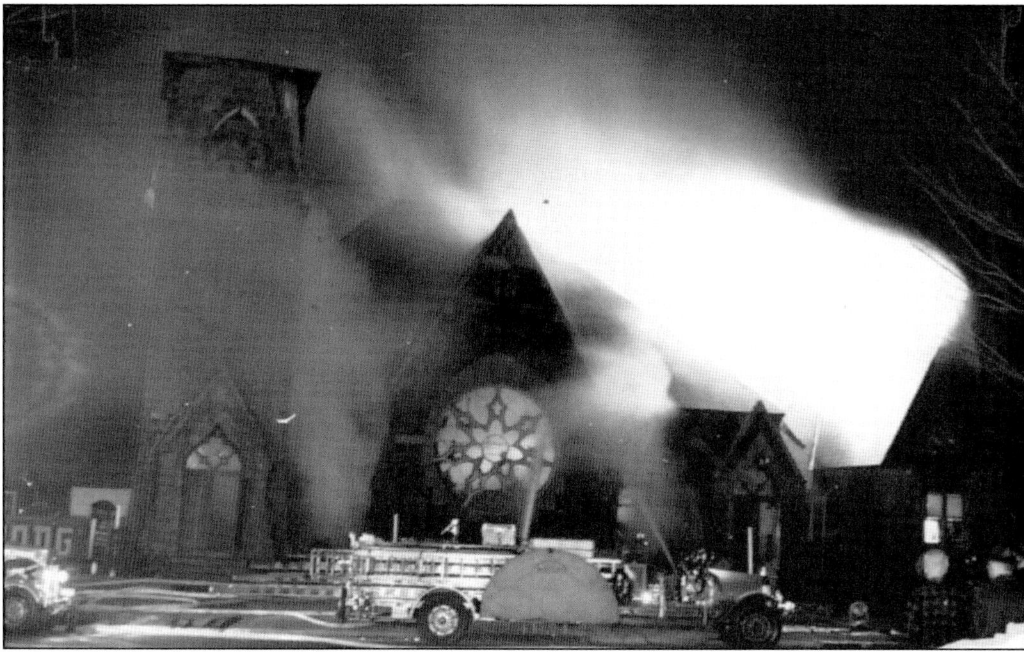

The department's 1926 Maxim ladder truck is working at the fire that destroyed the Congregational church on the corner of South Main Street and Farmington Avenue on January 3, 1942. The truck was sold to the Simsbury Fire Department in the 1950s and was retired in 1972. It was on display at the Connecticut Fire Museum until 1986. (Courtesy of the Rudy Desroches Collection.)

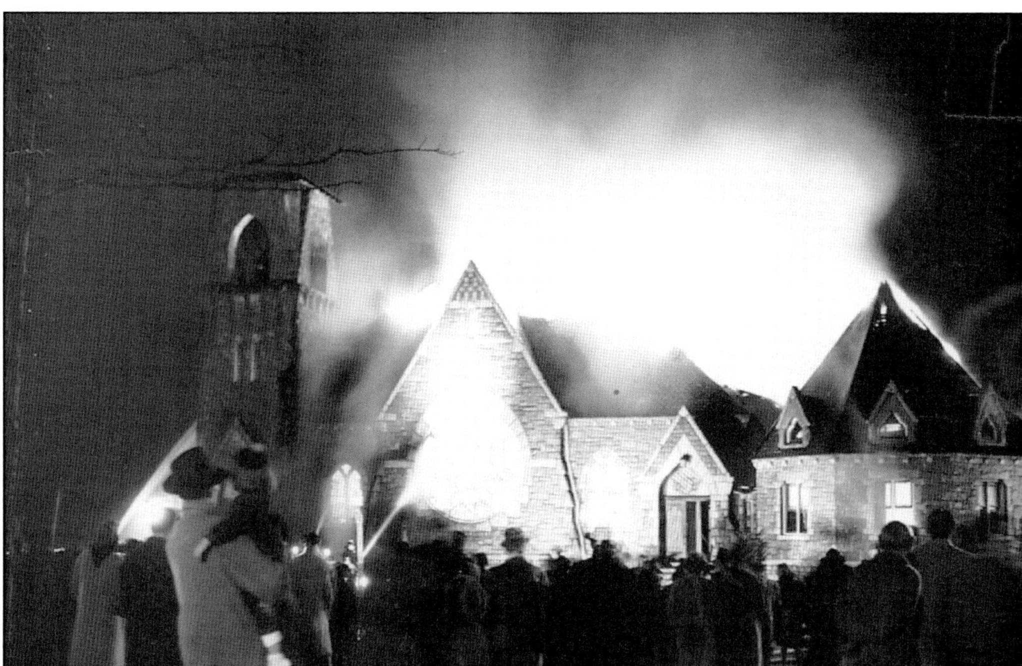

Firefighters are hosing down the fire that destroyed the Congregational church. (Courtesy of the Rudy Desroches Collection.)

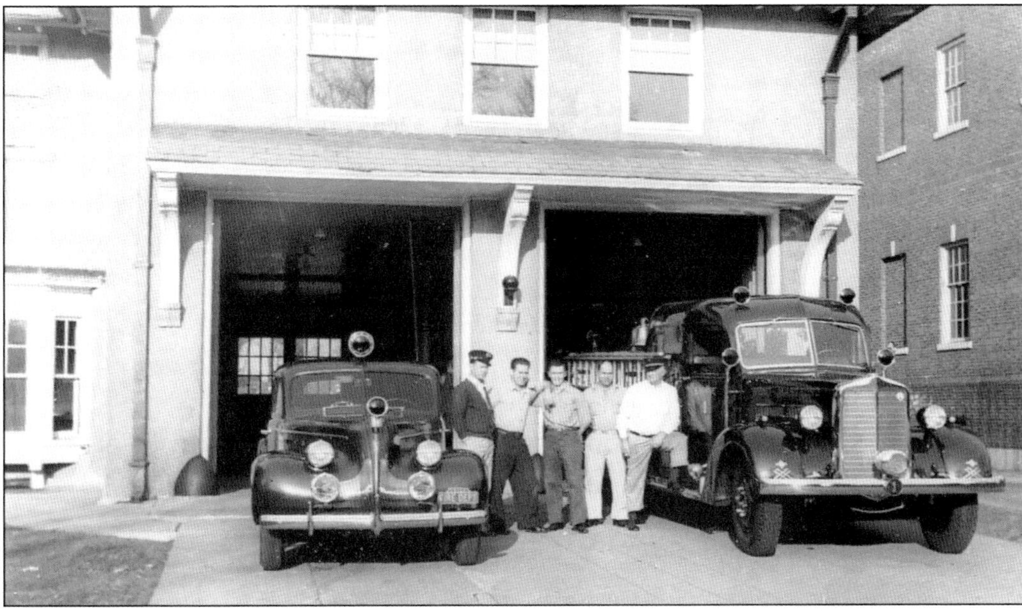

In this 1943 view are Engine No. 1's 1941 Maxim pumper and the chief's sedan waiting on the ramp. The chief or his deputy responded to all alarms in town. (Courtesy of the Rudy Desroches Collection.)

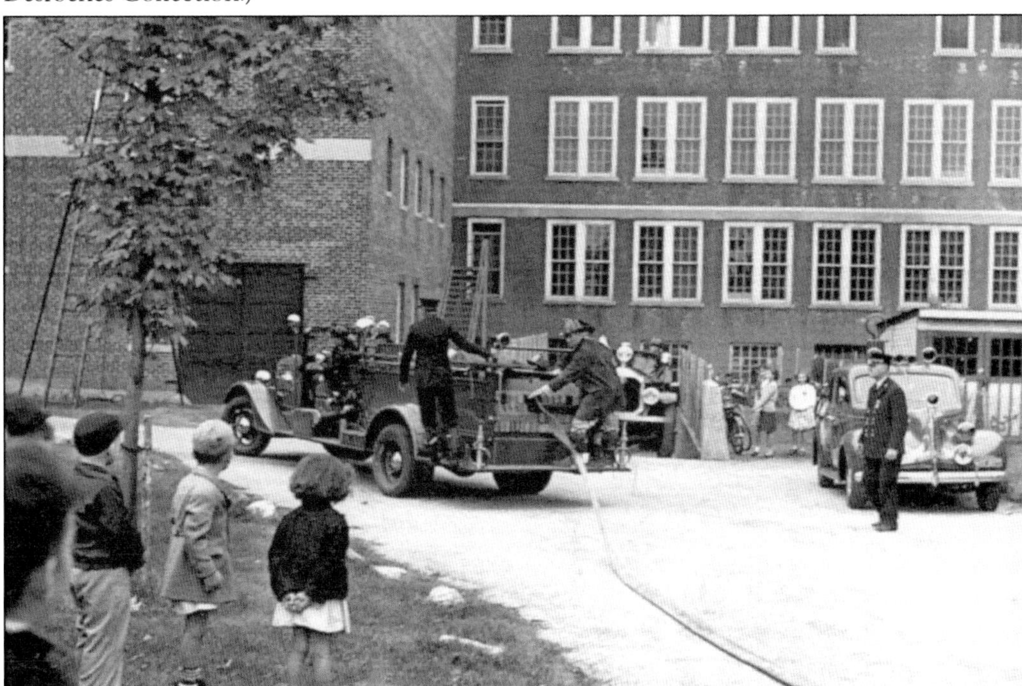

Engine Company No. 3's 1935 Diamond T pumper and hose car, along with the 1926 Maxim ladder truck, are giving a demonstration at the Plant School. In 1955, the Diamond T pumper was sold to the East Farmington Fire Department and then again in 1960 to the Boy Scouts of America for fire protection at the Lake of Isles scout reservation in Ledyard, Connecticut. (Courtesy of the Rudy Desroches Collection.)

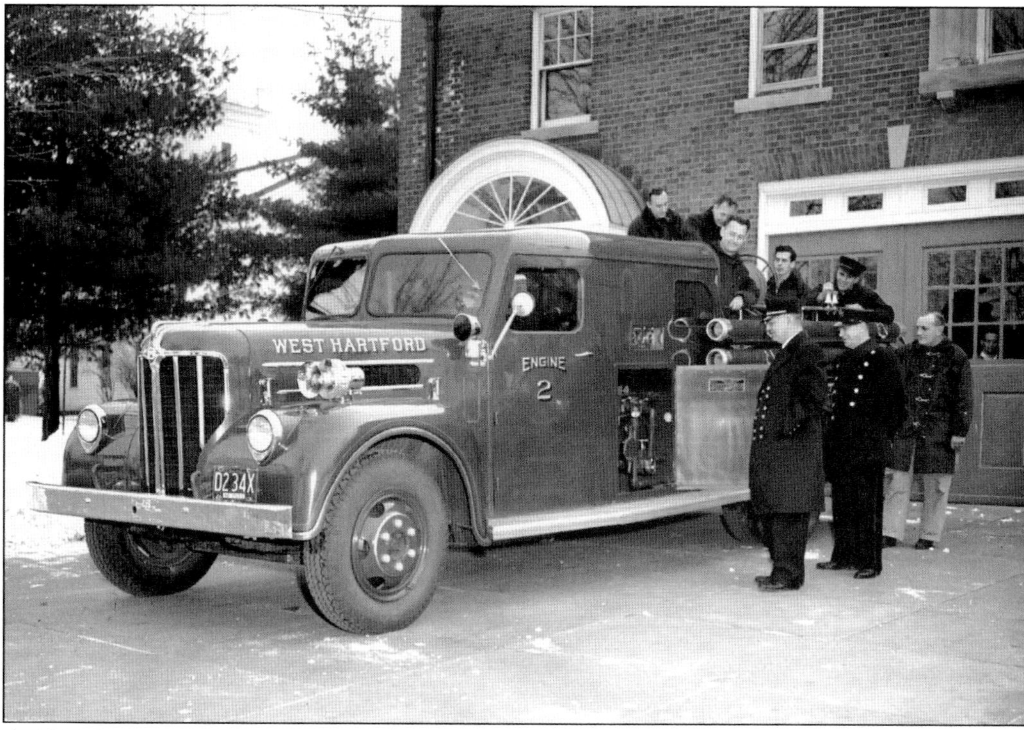

A 1948 Maxim pumper is seen in front of Station No. 2 shortly after being driven from the Maxim plant in Middleboro, Massachusetts. Note the Massachusetts dealer license plate is still on the truck. (Courtesy of the Rudy Desroches Collection.)

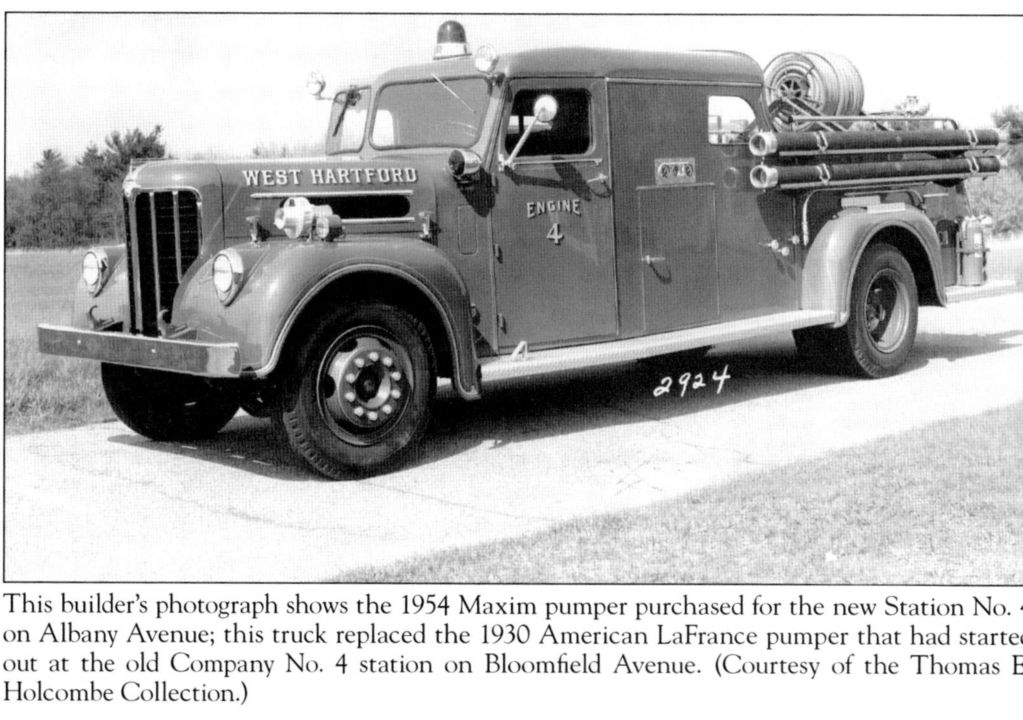

This builder's photograph shows the 1954 Maxim pumper purchased for the new Station No. 4 on Albany Avenue; this truck replaced the 1930 American LaFrance pumper that had started out at the old Company No. 4 station on Bloomfield Avenue. (Courtesy of the Thomas E. Holcombe Collection.)

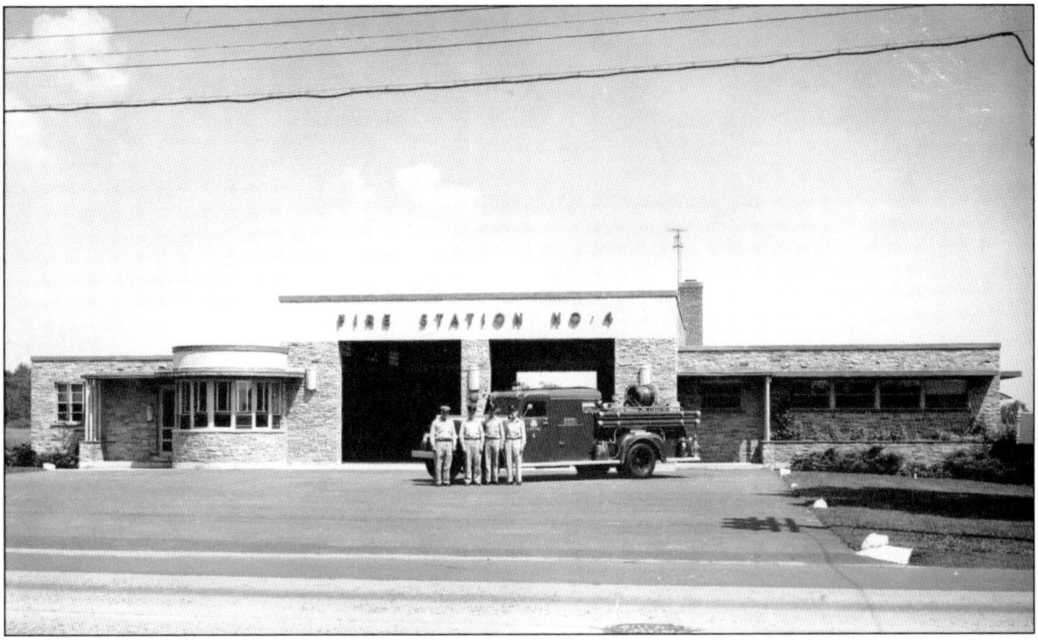

Newly built Station No. 4 on Albany Avenue is seen in 1954 with the new Maxim pumper. (Courtesy of the Rudy Desroches Collection.)

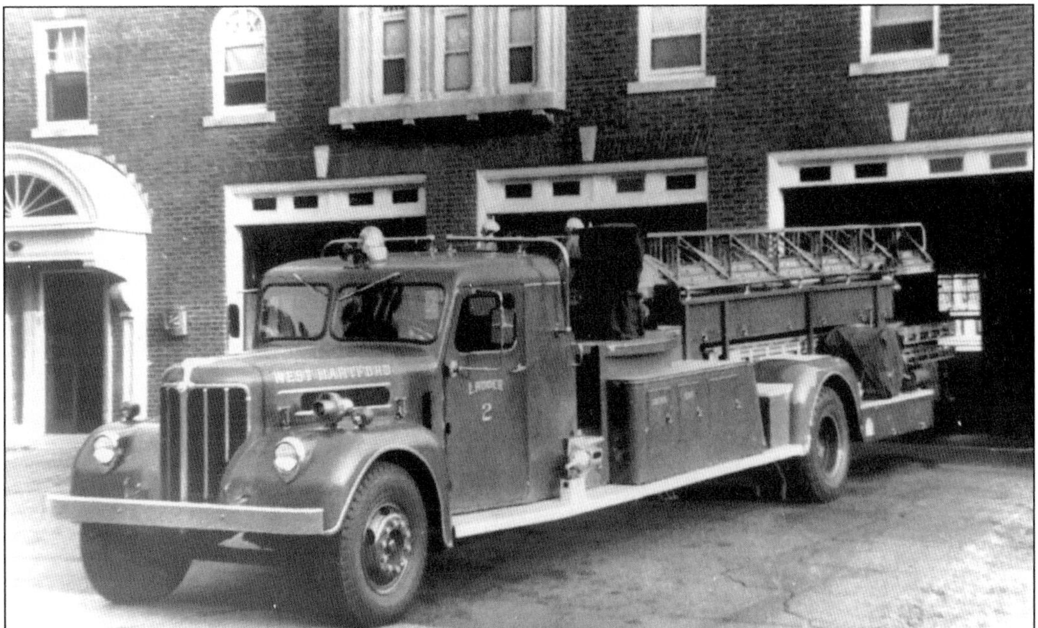

Seen here is a photograph of West Hartford's 1952 Maxim 75-foot aerial ladder truck in front of Station No. 2 in 1973. This unit normally operated out of Station No. 3 on New Britain Avenue as Ladder No. 1 but is filling in for Ladder No. 2, which was in the shop for repairs. (Courtesy of the Thomas E. Holcombe Collection.)

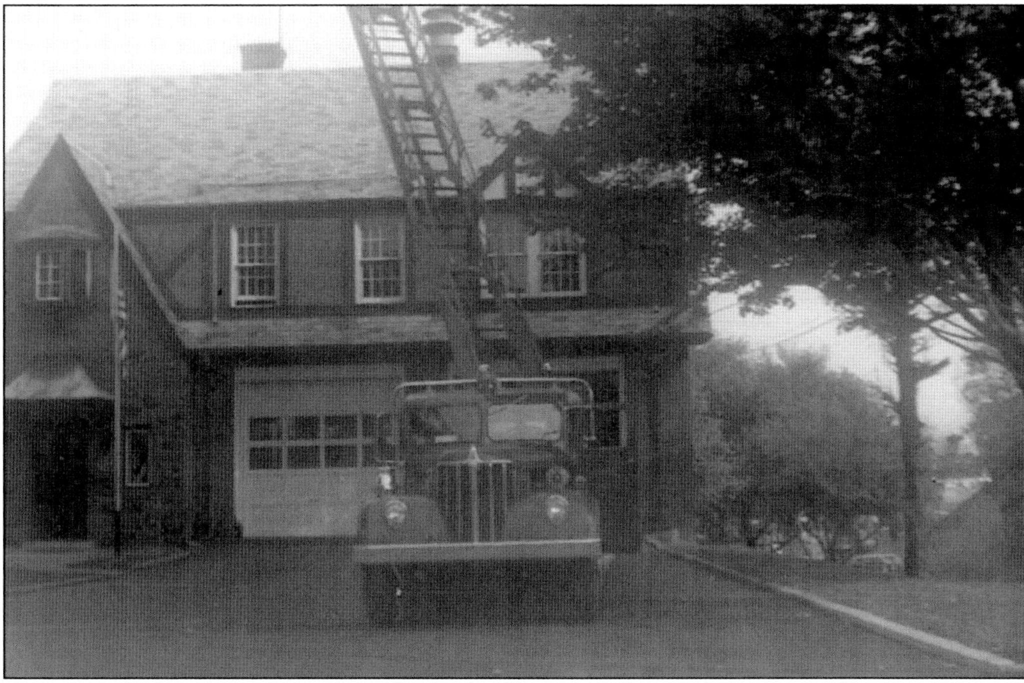

Seen here is another view of the 1952 Maxim ladder truck at Station No. 3. In 1974, it was replaced by a new Pirsch ladder truck and sold. (Photograph by Thomas E. Holcombe.)

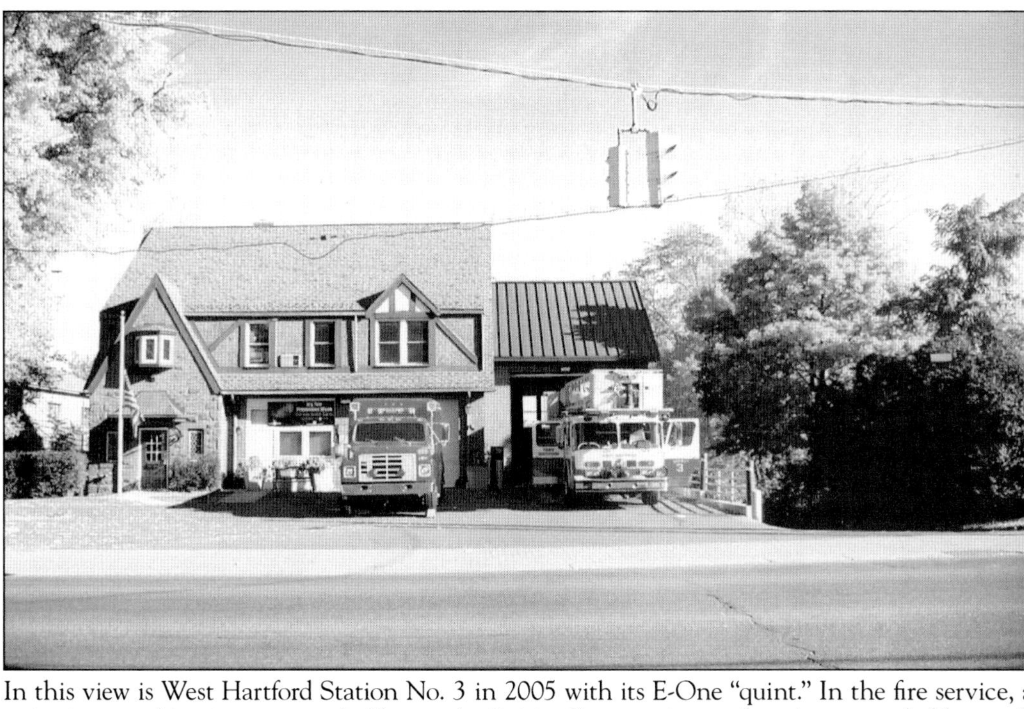

In this view is West Hartford Station No. 3 in 2005 with its E-One "quint." In the fire service, a quint is a combination pumper ladder truck. Originally, two pieces of equipment, a ladder truck and a pumper, operated out of this station. (Courtesy of the Connecticut Fire Museum Archives.)

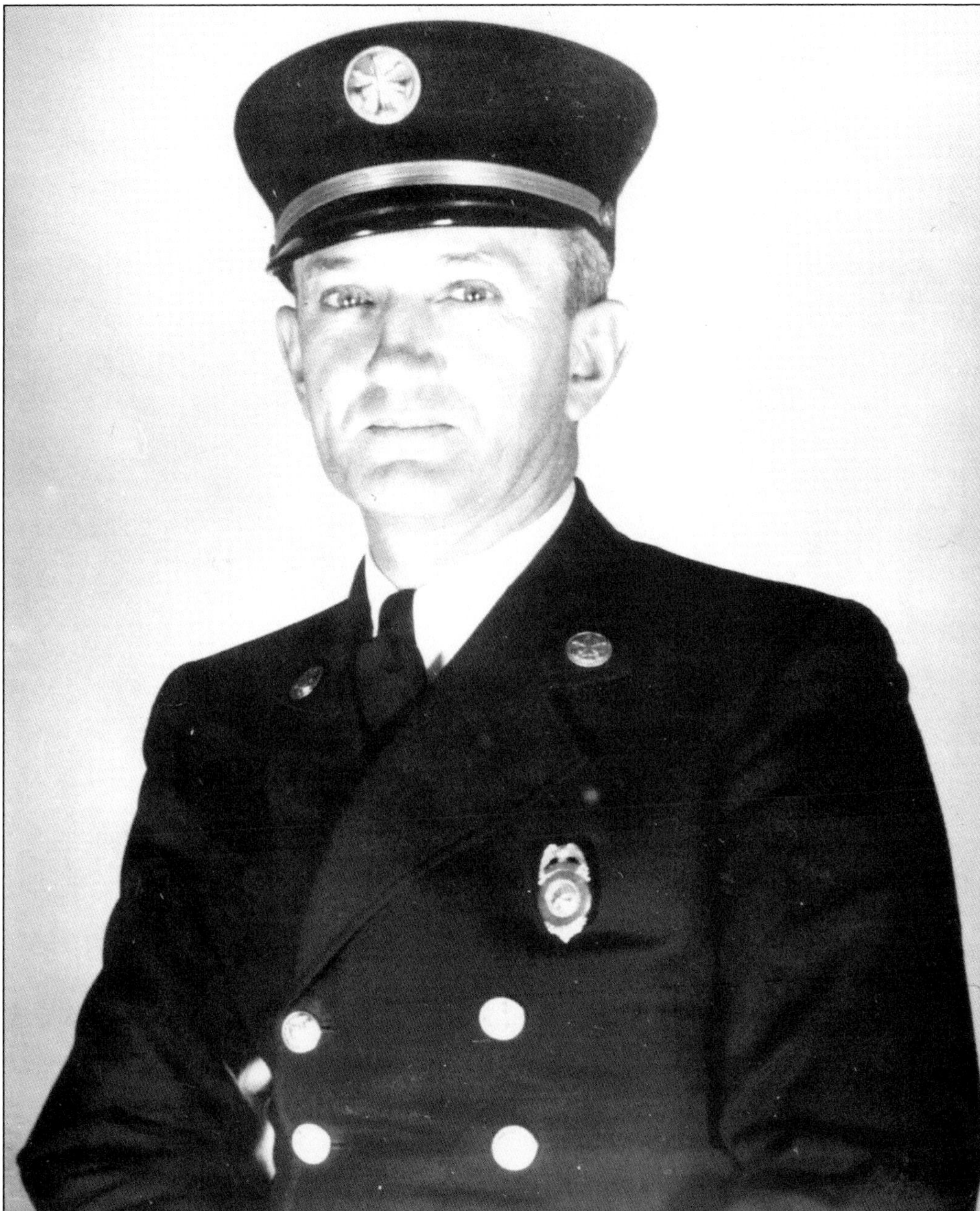

John T. O'Laughlin served as chief of the West Hartford Fire Department from 1956 until his retirement in 1967. He joined the department in 1930. (Courtesy of the Rudy Desroches Collection.)

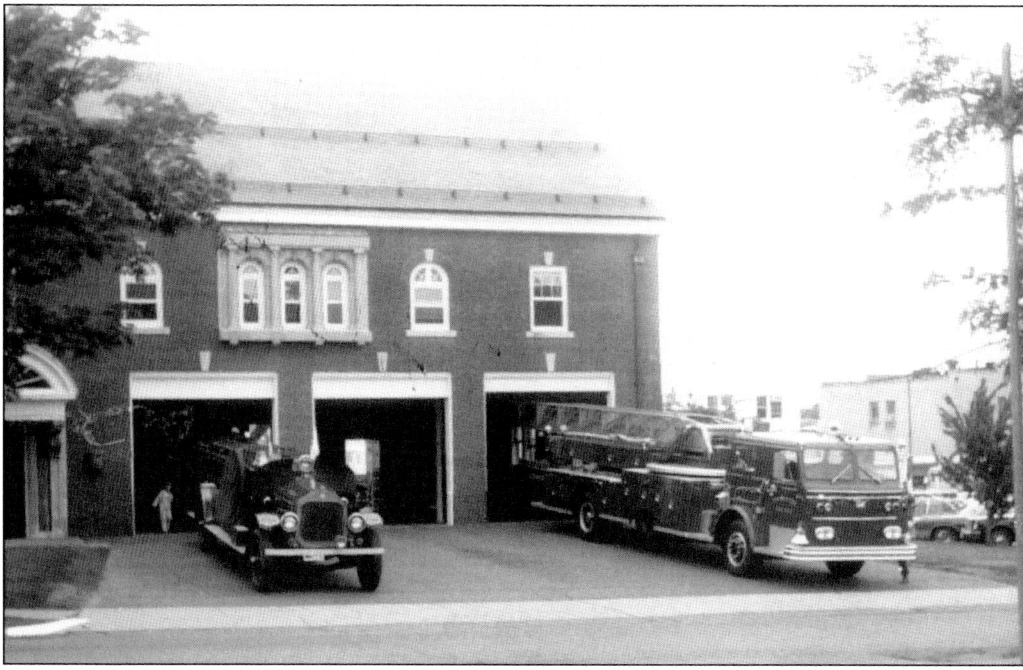

The Connecticut Fire Museum's 1926 Maxim ladder truck, formally Ladder No. 1 in West Hartford, is seen at its old station on Brace Road with West Hartford Fire Department Ladder No. 2, a 1967 Maxim, May 1976. (Photograph by Thomas E. Holcombe.)

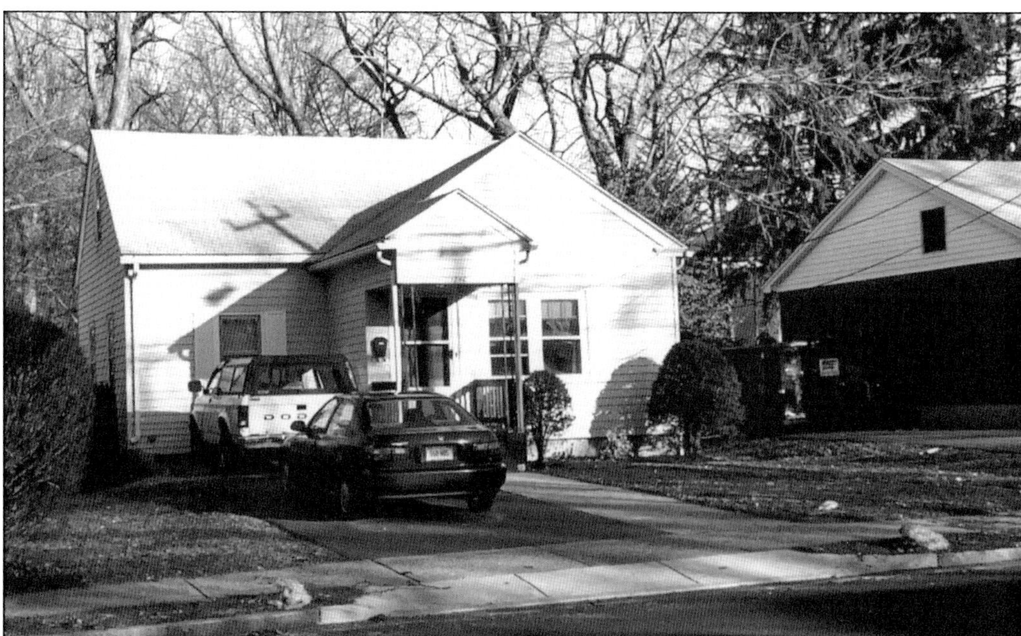

The former Quaker Hose Company station at 1346 Boulevard, now a private home, is seen in 2006. (Courtesy of the Connecticut Fire Museum Archives.)

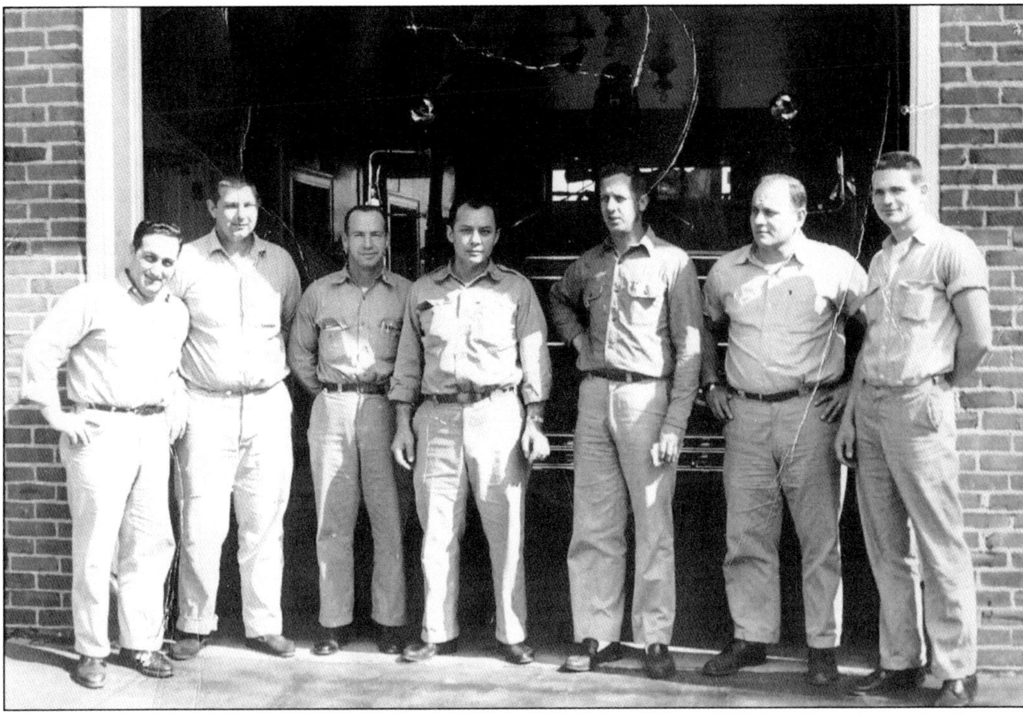

Station No. 3's crew is shown in this 1962 photograph. From left to right, they are Carmen Bruno, Richard Boulanger, Lt. Martin McMahon, Capt. John Kulow, Ralph Corning, Fred Bilak, and Jack Coppinger. (Courtesy of the Rudy Desroches Collection.)

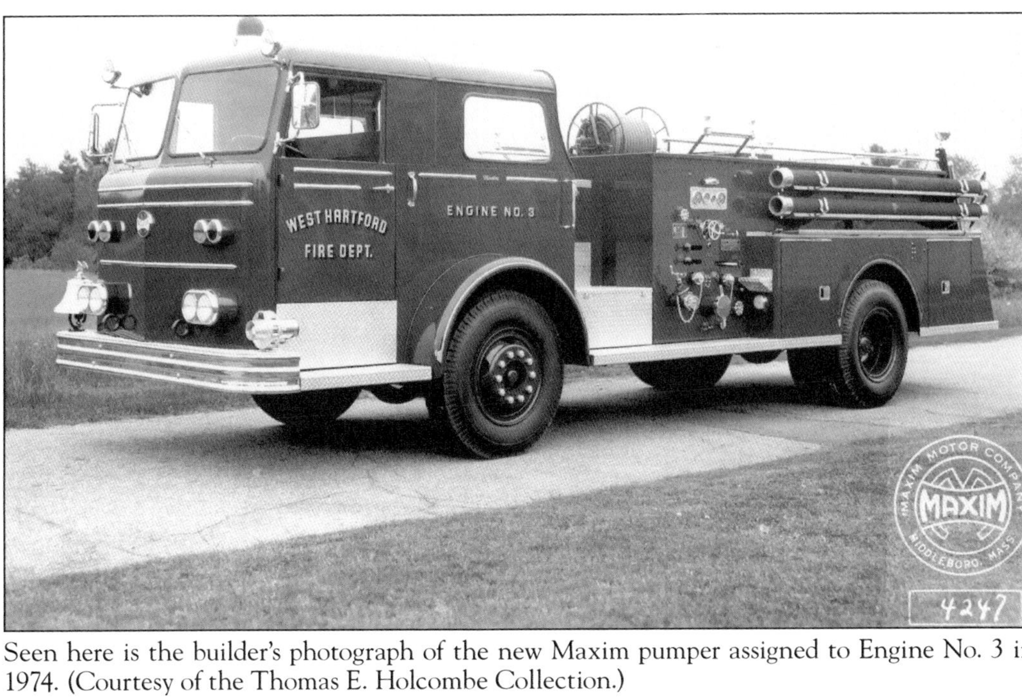

Seen here is the builder's photograph of the new Maxim pumper assigned to Engine No. 3 in 1974. (Courtesy of the Thomas E. Holcombe Collection.)

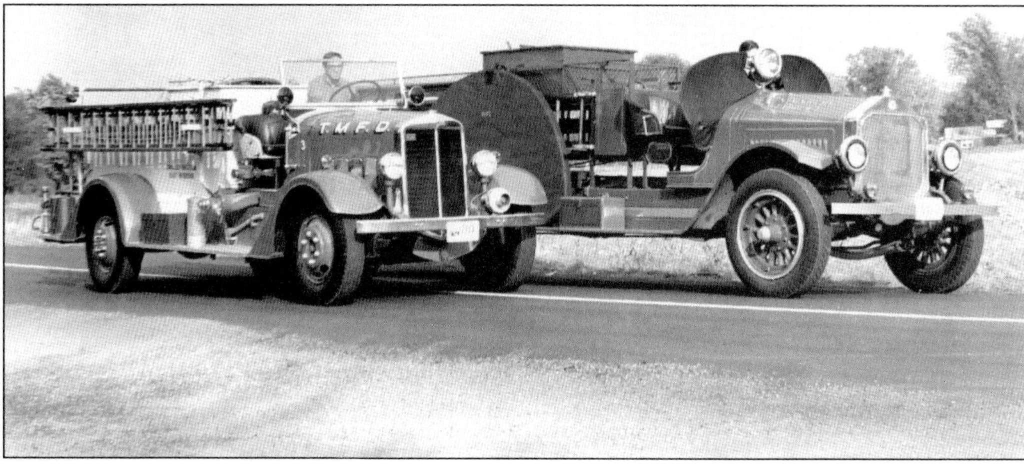

West Hartford's 1937 Maxim pumper, which served as Engine No. 2, and the 1926 Maxim ladder truck, which served as Ladder No. 1, were both on display at the Connecticut Fire Museum from 1973 to 1986. Both vehicles are now in private collections. Both served other towns after leaving West Hartford; the 1937 Maxim served Torrington, and the 1926 Maxim served in Simsbury. (Courtesy of the Connecticut Fire Museum Archives, photograph by Joan Hogaboom.)

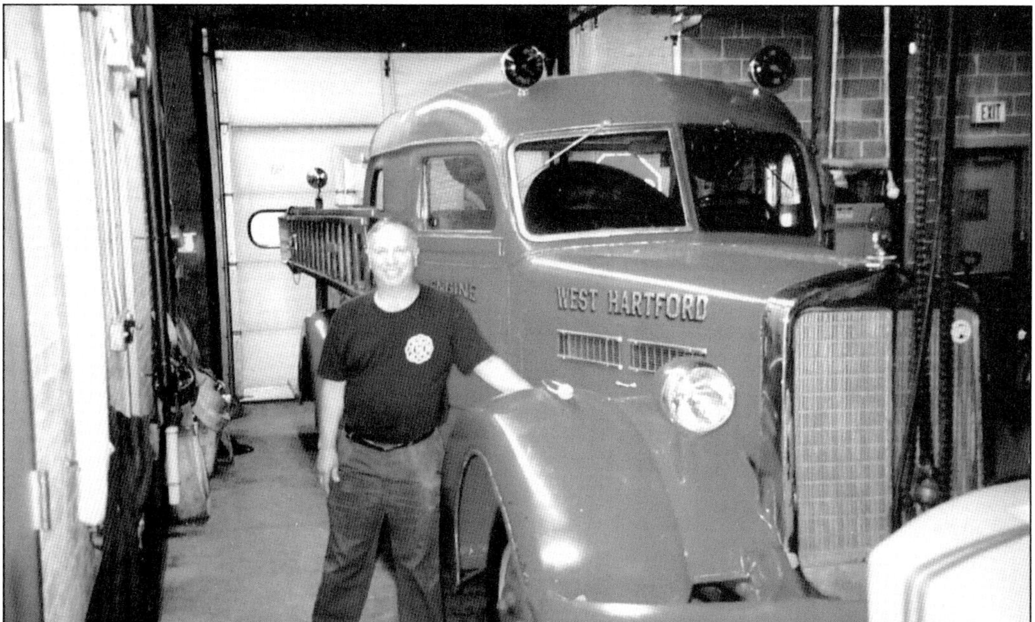

Raised in West Hartford, Thomas Holcombe became interested in fire apparatus at the age of five. In the 1950s and 1960s, he was an avid fire engine "chaser" and became very familiar with all the apparatus of the department. In 1973, he joined the Connecticut Fire Museum and has spearheaded its restoration efforts for the past 25 years. His extensive knowledge on the mechanical operation of fire apparatus has been an invaluable resource for the museum. Holcombe lives in Burlington, Connecticut, with his wife, Ava. He is shown here with the department's 1941 Maxim pumper that was recently purchased from an individual in Alabama. The department has plans to restore the vehicle for use at parades and exhibitions in the area. (Courtesy of the Connecticut Fire Museum Archives.)